P9-DWY-350

BIOLOGY, ETHICS, AND ANIMALS

179.3
R612

BIOLOGY, ETHICS, AND ANIMALS

ROSEMARY RODD

WITHDRAWN

CLARENDON PRESS · OXFORD

LIBRARY ST. MARY'S COLLEGE
193164

Oxford University Press, Walton Street, Oxford OX2 6DP

Oxford New York Toronto
Delhi Bombay Calcutta Madras Karachi
Petaling Jaya Singapore Hong Kong Tokyo
Nairobi Dar es Salaam Cape Town
Melbourne Auckland

and associated companies in
Berlin Ibadan

Oxford is a trade mark of Oxford University Press

Published in the United States
by Oxford University Press, New York

© Rosemary Rodd 1990

First published 1990
First issued in paperback 1992

All rights reserved. No part of this publication may be reproduced,
stored in a retrieval system, or transmitted, in any form or by any means,
electronic, mechanical, photocopying, recording, or otherwise, without
the prior permission of Oxford University Press

British Library Cataloguing in Publication Data
Data available

Library of Congress Cataloging in Publication Data
Rodd, Rosemary.
Biology, ethics, and animals / Rosemary Rodd.
1. Animal welfare—Moral and ethical aspects.   2. Animal rights.
3. Bioethics.   I. Title.
HV4708.R63   1990   179'.3—dc20   89–70902
ISBN 0–19–824052–X

Printed in Great Britain by
Bookcraft (Bath) Ltd,
Midsomer Norton, Avon

$18.95 WJW 3-1-93 (28)

LIBRARY ST. MARY'S COLLEGE

ACKNOWLEDGEMENTS

Many people have given generously of their time to provide helpful criticism and suggestions for this book. I would particularly like to thank my supervisors, Tom Sorell and Jenny Teichman, who steered me through the production of the thesis upon which it is based, and Mary Midgley, who examined the thesis and subsequently offered many helpful suggestions for the transformation into book form. Gill Langley of the Dr Hadwen Trust for Humane Research kindly read through the first draft of the section on alternatives to the use of animals in scientific research. However, any errors and omissions which remain are, of course, my responsibility. The opinions expressed in the book are personal views and are not to be attributed to any organization with which I am, or have been, associated.

R.R.

CONTENTS

1

BIOLOGY AND VALUES

This work is a philosophical examination of the significance of theories and factual discoveries from the life sciences for the development of ideas about the moral standing of animals. Individual biologists whose work is discussed will usually be mainstream researchers whose contributions are broadly acceptable to their peers (although the latter might not necessarily agree with all their conclusions).[1] Where this is not the case this will be indicated in the text. I hope that the work will be of interest both to biologists who wish to explore the implications of their science, and to philosophers and general readers. One of my main aims (particularly in the second chapter) is to provide sufficient background knowledge to help the latter to discuss and evaluate in an informed way the things which biologists say about animals. A particular theme will be the need to consider how decisions should be made about the way in which different kinds of species ought to be treated, with especial reference to attempts to reduce the suffering caused in biomedical experiments by changing the types of organisms used as subjects.

I should point out that this work is not intended to be a comprehensive discussion of the theory of animal rights, and that it is intended to be read in conjunction with more broadly philosophical treatments of the subject. A suitable comparison may perhaps be found with books discussing practical issues of medical ethics, such as Raanan Gillon's *Philosophical Medical Ethics*.[2] Gillon does not think that doctors need to be convinced that they owe a moral duty to their patients, but that they want help in deciding what to do about various ethical dilemmas they are likely to face in the course of their medical practice. Readers looking for a detailed examination of

---

[1] Citation by other authors working in the field of biology has been taken as the main indication of this. [2] (Chichester: Wiley, 1986.)

the theory of rights or for detailed lists of what we may or may not do to animals are likely to be disappointed. However, I hope they will find that the book provides some helpful guidelines on how we should go about assessing our behaviour towards non-human animals in various situations, and goes some way to providing an affirmative answer to the question 'Is it biologically possible for us to experience a genuine moral concern about our fellow creatures' welfare?'

The idea that values or 'ought' statements may be derived from purely factual evidence has been in general disrepute since the time of Hume; thus it would seem that we cannot discover that we have obligations to animals (or to other people) from facts about their nature: that they can feel pain etc. However, I think we have to admit that some kinds of fact necessarily include a notion of value: notably facts about mental states. The fact that a mental state is an unpleasant mental state also implies that the state is a bad and undesirable one. Conversely, the fact that a mental state is pleasant implies *ceteris paribus* that it is good and desirable. This does not simply represent a matter of taste: it is logically possible to imagine someone who enjoys painful feelings, but then these have to be described as pleasant. (For example, in *A Midsummer Night's Dream*, Helena likes the idea of being beaten up by Demetrius, but is terrified at the idea of being attacked by Hermia. It is perfectly possible for us to describe her feelings in the first instance as 'pleasant' and in the second as 'unpleasant' and this 'higher-level' description fits quite happily into a schema in which conscious agents want pleasant states and don't want unpleasant ones.) Similarly, early Christians despised the pursuit of pleasure, but it would be quite false to say that they actually *valued* unpleasantness for its own sake. It makes much more sense to say that they had different ideas about what constitutes happiness, and that they believed that they should place a greater value upon eternal happiness than upon immediate, worldly pleasure. Thus it is a matter of taste whether certain *things* are good or not (I may like coffee, my cat likes the brand of food 9 out of 10 owners said their cats preferred, my rabbit would rather have a lettuce), but on a secondary level of description the mental states I, the cat, and the rabbit have when we get coffee, brand X, or lettuce are similarly desirable states. In isolation, coffee, brand X, and lettuce are merely neutral. Thus it is reasonable to draw up a premiss: mental states can possess positive value (good) or negative value (bad). Then, if we want to increase good and to decrease evil we ought to consider in our calculations all those beings which are likely to possess mental states. Purely descriptive factual evidence can then make it possible to decide what kinds of beings are likely to possess mental states, and what factors are likely to affect these states. Of course there are potential complications about mental states which are bad but enjoyed by the person who is experiencing them (for example, someone who enjoys causing pain to others), but in principle I don't think it is impossible to say that here we would consider the situation as a whole and still judge that the sadist's enjoyment had minus value. Similarly, there is no bar to a ranking order which grades pleasant and unpleasant states in some way by quality (enjoyment of Mozart vs. Match of the Day), as well as by intensity.

I shall not attempt to produce an argument which could show why we should want to increase good, except to suggest that the connection between 'good' and 'ought to do' is of the same type as the relation between 'poisonous' and 'ought not to be eaten'.[3] That something is poisonous is a reason not to eat it, but, as Hare says, there is not a logical connection between the two sentences 'This is poisonous' and 'This ought not to be eaten'. Because of these things, I believe that there is something useful to be gained from an exploration of present knowledge about the natures of non-human animals in order to clarify the reasons for thinking that some of them ought to be treated as creatures with moral status. I also hope to show how thought about the natures of humans and non-humans has been coloured by the philosophical attitudes current in the biological sciences, and what this implies for philosophers interested in the question of the moral status of animals.

Some authors, notably R. G. Frey in his book *Interests and Rights: The Case Against Animals*,[4] have suggested, that, if it

---

[3] R. M. Hare, 'Descriptivism', in W. D. Hudson (ed.), *The Is–Ought Question* (London: Macmillan, 1969), 257–8.

[4] (Oxford: Clarendon Press, 1980.) Frey is himself dubious about the value of the concept of rights in general.

is valid to talk about rights, then these must depend upon the ability to possess interests, which in turn depends upon the ability to form verbal concepts, specifically, verbal formulations of desires and claims of rights. This would rule out the possibility of rights for non-human animals, with perhaps the controversial exception of some of the great apes and cetaceans, but would leave open the possibility that beings incapable of possessing genuine rights might possess moral status in virtue of other qualities, such as the capacity for suffering.

Throughout this book, I shall generally take a correlative view of rights: that is, that if it is the case that we have duties to an animal, then that animal has corresponding rights. It seems reasonable to distinguish duties in respect of individual sources of consciousness (such as humans and sentient animals) from duties in respect of non-conscious objects (such as the Mona Lisa or a giant redwood), which seem to be duties *about* rather than duties *to*, and appear to be owed to other conscious beings who are liable to benefit from them; or perhaps even to ourselves (because of the bad effects on us of habits of wanton destruction). Hence, we have duties to babies and cows, which means that we assume that they are kinds of creatures who can have rights; while we have duties about trees (such as a duty not to exterminate them) which entail no assumption that they have rights. Thus, sentient animals and infant humans would be grouped with human adults, rather than with valuable inanimate objects.

However, I would not object too much to an alternative view which grants moral status to all sentient creatures while reserving 'rights' purely as a technical term to describe valid claims which are the result of verbal agreements between moral agents. On this view, all humans would possess a basic moral status which might be augmented by reciprocal agreements. Thus, for example, if I lend one of my books and the borrower promises to return it, I can be said to have a right to repossess the book. My basic moral status as a sentient being would determine the answer to different questions, such as the reason why it would be wrong to cause me to starve to death.

A difficulty with this concept of rights is that it seems to make them rather exclusive. If rights are valid claims based upon verbal agreements, then not only those creatures incapable of formulating such agreements, but also certain classes of normally intelligent adult humans may be removed from the class of rights-holders, or may at least come to have drastically fewer rights than the rest of us. Notoriously unreliable people, for example, would tend to become excluded from the network of rights-granting agreements (because it would not be in the interest of trustworthy people to make contracts and bargains with them). It appears to me that we should then be compelled to say that only good people can have genuine rights.

Similarly, I shall not attempt a detailed defence of the concept of rights against a view that they are superfluous if the importance of moral principles (such as avoiding cruel actions) is taken seriously. I shall take the view that to claim that a particular kind of being is a rights-holder is a significant stand because it implies that we ought to adopt a particular attitude towards that class of entities, but that it is essentially a polemic, rather than factual statement. It is intended to point up a contrast with non-rights-holders, which may have some kinds of moral worth, but in respect of which we have a much more limited set of obligations. (For example, our obligations with respect to the Mona Lisa are not likely to encompass more than a duty not to destroy it, and perhaps a duty not to make it inaccessible to other people; our duties towards an animal are likely to imply much more generalized concern to avoid causing adverse mental states.)

I shall not devote much space to a discussion of the possible taxonomy of rights of different kinds (such as liberty rights, rights of participation, welfare rights, special rights, and so on). However, at appropriate points I hope to leave indications of how it would be possible to show that members of at least some species of non-human animals would be plausible candidates for the different types of right which have been postulated for humans.

I shall generally take the view (against Frey)[5] that, where animals are conscious at all, they experience emotions, rather than merely pleasant or unpleasant mental states, and that they are capable of simple desires. Frey argues that both types of experience depend upon language and self-consciousness,

---

[5] *Interests and Rights*, pp. 104 ff., 121 ff.

but I think that his analysis is faulty. Suppose that we follow his example of a dog rushing towards a bone. Frey claims that we must say either that the dog simply desires the bone but is unaware of simply desiring it or that the dog simply desires the bone and is aware of simply desiring it. If we claim the former, then, according to Frey, the concept of a desire becomes devoid of significance. If the dog is unaware of desiring the bone, we must say that this is an unconscious desire, but if the desire is unconscious Frey claims that we cannot hold that it genuinely differs from (for example) a plant's non-conscious tendency to grow towards the light. If we claim the latter, we have to say that the dog is self-conscious, which Frey believes is not possible without language. The questions of language and self-consciousness will be discussed at length in Chapters 3 and 4, and I am here concerned only to refute Frey's rejection of the concept of simple desires. I think that Frey's conclusion depends largely upon the formula he uses: 'The dog simply desires the bone but is unaware that it simply desires the bone'. If the proposition is reworded as 'The dog simply desires the bone but is unable to reflect about her desire', it seems to me that we no longer feel compelled to say that the dog's desire is in any way necessarily unconscious. The dog feels the desire, but is not able to stand back mentally and see herself desiring the bone. The plant, on the other hand, doesn't feel anything. Human self-consciousness is a capacity to reflect about desires and avoid acting merely upon the emotion of the moment. Unconscious human desires are something quite different and (at least in the Freudian sense) generally pathological results of inability to face up to ourselves.

I shall also hold that to say that x is in A's interest need not imply more than that x would contribute to A's good and that A is a being capable of possessing mental states. There clearly is a difference between allowing a dog to die of thirst (result: material damage plus very unpleasant mental states) and failing to water a pot plant (result: only material damage).

It appears that value can plausibly be said to reside in certain mental states of conscious beings. When we talk about the interests of conscious beings we point to what will enable them to achieve states which have positive value. We also recognize that all conscious beings will tend to strive to achieve

positive value for themselves and to avoid states which have
negative value. Thus there is a connection between the interests
of conscious beings and the sources of value. If we want to
promote what is good, we have to address ourselves to the
interests of conscious beings.

Broadly speaking, an individual's basic needs are those factors
without which she will either be unable to achieve a net positive balance of mental states or be unable to continue to survive
Clearly, we cannot be expected to devote our lives to maximizing the positive value of other individuals' mental states: that
someone might be happier if she had a larger car is not a
compelling reason why I should act to satisfy that particular
interest. However, it seems to me that it is reasonable to say
that we ought to be concerned whenever we are liable to interfere with basic needs, i.e. to reduce the value of an entity's
mental states to zero or less. Furthermore, if we think of the
rights of others as that which is owed to them, then, if what is
owed is a fair recognition of their own striving to achieve
personal value, rights must ultimately depend upon fairness in
consideration of basic needs.[6]

Questions about mind and consciousness in other animals
are of central significance for a theory of their moral status.
Here, ideas have been profoundly changed by new biological
theories. After Darwin, no philosopher would be likely, even
tongue-in-cheek, to argue like Voltaire:

What kind of man will dare to affirm, without absurd impiety, that it
is impossible for the Creator to endow matter with thought and
feeling? . . . The animals have the same organs as ourselves, the same
feelings, the same perceptions; they have memory, they put together a
few ideas. If God has been unable to animate matter and give it
feeling, you are left with one of two things: either the animals are pure
machines or they have a spiritual soul.

[6] See David Wiggins's paper, 'Claims of Need', in T. Honderich (ed.),
Morality and Objectivity (London: Routledge & Kegan Paul, 1985), 149–202,
for a discussion of the connection between obligations and needs. I do not
mean to imply that animals (or humans) always, or even generally, know what
is in their own best interest. It is possible to want something which will
ultimately harm our interests. However, I think it is then true to say that we
have made a mistake of judgement (like Mill's man who thinks that he wants
to cross a dangerous bridge, but is mistaken because it would lead to a fall into
the river, which he does *not* want ('On Liberty', in *Utilitarianism*, ed. Mary
Warnock (London and Glasgow: Collins, 1962), 228–9)).

It seems almost proven to me that the animals cannot be simple machines. Here is my proof: God has provided them with exactly the same organs of feeling as ours. So if they cannot feel, God has done a pointless job. Now by your own admission God does nothing in vain. Therefore He has not made all these organs of feeling so that there should be no feelings. Therefore the animals are not simple machines.

The animals, according to you, cannot have a spiritual soul, so, in spite of you, there remains nothing else to say except that God has given to the organs of animals, who are matter, the faculty of feeling and perceiving, which in them you call instinct.[7]

A modern Voltaire would be much more likely to point to the evolutionary continuity between humans and animals, and to the improbability of supposing that sense organs give rise to true sensation only in *Homo sapiens*.

The traffic between biological and ethical theorizing is not unidirectional. It is now widely accepted that scientific theories are not fully determined by factual evidence, but depend also upon general assumptions about the nature of the world and upon value judgements about the nature of a good theory.[8] The overriding value upon which theories are judged is that of successful prediction and control of external events, but scientists are influenced by other values, such as simplicity. In 'Theory and Value in the Social Sciences', Mary Hesse argues that values, in the sense of beliefs about the way society ought to be, have a legitimate part to play in determining sociological theories. Because of the complexities of human societies, and because it is not possible for the sociologist to act as an objective observer (since she is herself the product of human society), it will never be possible for the value of successful prediction to replace all others in deciding between one theory and another. Since this is so, Hesse suggests that it is better for us to make a conscious and explicit decision to choose values which will promote the welfare of society, rather than cling to a spurious ideal of objectivity which in fact surreptitiously allows the interest of the dominant power group to become the deciding

---

[7] *Letters on England*, trans. Leonard Tancock (Harmondsworth: Penguin, 1980), 66–7.

[8] Thomas S. Kuhn, *The Structure of Scientific Revolutions* (Chicago: University of Chicago Press, 1962).

factor by default.[9] Similarly, I believe that we should accept that, where pragmatic criteria are insufficient to decide between biological theories—for example, to decide whether an animal reacts to a stimulus because of conscious perception, or merely mechanically—we should adopt the principle that animals ought not to be described as feelingless automata unless there is positive factual evidence that this is so. This principle cannot be defended upon pragmatic grounds, but there are other reasons why it should be favoured, for example, because describing animals as automata when they are not may lead us to inflict needless pain upon them and because the world would be a poorer place if humans were the only conscious elements within it. As Hesse argues in the case of human sociology, if we adopt a spurious posture of objectivity we are liable to choose the theory which favours the interest of the dominant group—here the theory that humans are free to use animals in any way they wish, because animals can have no feelings. We need to make positive selections of theory-determining values instead of pretending that we have no bias.

Work done in the philosophy of science also has very practical significance for considerations of the extent to which it is possible to find substitutes for living animals in scientific research. Part of the argument about the possibility of replacing animals concerns the nature and function of model systems in the development of scientific theories; the extent to which observed facts are constrained by the use of theory-laden descriptive language; and the relationship between fact and theory. Study of the way in which scientific theories are developed can help us to go beyond a simplistic view that medical science advances by amassing facts which are gained by trying out possible treatments on animals to a more realistic investigation of what constitutes a good model of the diseased state and how this relates to the pragmatic search for successful treatments.

A major theme of this book is the way in which certain biological theories about the evolution of human and animal behaviour have affected the attitudes of biologists to moral arguments about the status of animals. Borrowing Raanan Gillon's

---

[9] 'Theory and Value in the Social Sciences', in C. Hookway and P. Pettit (eds.), *Action and Interpretation: Studies in the Philosophy of Social Interpretation* (Cambridge: Cambridge University Press, 1978), 1–16.

technique (as he says, paraphrasing thoughts which are more
often expressed in conversation than in print),[10] the views of a
typical biologist might be set out as follows:

There is something very unnatural about worrying too much about
what we do to other species. We are natural predators, and it is natural
for us to exploit animals for our own benefit. Besides, look how much
farm and laboratory animals gain by being useful to us—all species of
domesticated animals are vastly more numerous than their counter-
parts in the wild, and this means they are vastly more successful in
evolutionary terms.

Human behaviour, including ethical behaviour, has only developed
because it had survival value. In other words, disapproving of particular kinds of action tended to increase the chance of the disapprover's
genes becoming more frequent in the gene pool. So it's obvious that
ethics is only relative anyhow—if natural selection had favoured
sadists we would think torture was admirable behaviour. We have a
tendency to disapprove of cruel behaviour because it is beneficial to us
to stamp out cruel tendencies (e.g. in children) before they can be
directed towards ourselves, not because there is any absolute scale of
values by which we can judge cruelty to be wrong. If we cause animals
to suffer because that is the best way to achieve some useful goal
(such as raising more or cheaper food, or testing medicines), it is a
mistake to say we are cruel since the suffering is only a by-product,
not something we enjoy. So the whole animal rights movement is a
misguided attempt to do something nature never intended.

I hasten to say that not all biologists believe anything like this,
and that those who do genuinely think it is the correct interpretation of the facts about human evolution. A very brief summary
of modern evolutionary theory will provide enough background
knowledge to understand the discussion of these, and related,
ideas which will be developed later in the book:

Evolution occurs through changes in the frequencies of variant
gene forms within populations of organisms. The totality of
genes within an interbreeding population is referred to as the
gene pool. New variant forms are continually, but slowly,
added to the pool by the process of random mutation—small
changes in the coding sequence of the DNA molecules due to
chance events, such as miscopying during cell division, damage
by radiation, or infection by viruses. Some variant forms may

---

[10] *Philosophical Medical Ethics*, pp. 28–9.

have characteristics which tend to increase their frequency within the gene pool: for example, a gene which causes its bearer to share food with siblings may tend to increase because siblings have a 50 per cent probability of possessing any gene in common; hence the gene is, in effect, causing copies of itself to be fed. It is important to remember that these kinds of selective effects do not necessarily depend directly upon success in sexual reproduction: for example, worker bees are sterile, but their genes remain frequent because they are carried into the next generation by the fertile queen and drones, which have the same genes as workers and are 'switched' into the reproductive form by environmental factors. It must also be remembered that, although genetic events may cause purposeful (i.e. goal-orientated) behaviour at the macroscopic level, the underlying changes in frequencies are entirely mindless physical and chemical processes no different in principle from the changes in frequencies of chemical molecules which occur when oxygen and metallic iron react together and produce the more stable compound, iron oxide. Here, one might loosely say that the molecules react together in order to produce a more stable state, but clearly no sense of purpose is intended. Again at the level of genes, change simply takes place in a way which leads to more stable gene forms, even though the mechanism of this stability may be more complicated than those involved in simple chemical reactions. In fact, some genes may do nothing besides promoting their own replication. It appears that their sole action is an increased tendency to appear in the fertilized embryo, relative to their alternative forms, and that they confer no benefit upon their macroscopic 'host', or are even mildly deleterious to it. The laws of population genetics are central to an understanding of biology, but they are not in principle any more like human laws (i.e. rules of society) than the gas laws which govern the mechanisms by which we breathe. Like the gas laws, they are statistical in nature, and a particular variant need only possess a greater probability of appearing in the next generation of a given population than the alternatives, if its frequency is to increase in the gene pool. Thus, for example, a gene for longer legs might on balance be an advantage (and so increase), even if it, on rarer occasions, turns out to be disadvantageous or even fatal (for example, it might increase the

probability of breaking a leg). The coupling between behaviour
and genetic success is relatively loose: a particular behaviour
pattern need only be statistically more successful than the
alternatives to become established in a population, and it is
possible for two or more alternative possibilities to coexist
(polymorphism) if both of them are fairly successful some of
the time.

These are the basic facts about evolution, and I think it is
important that they should be clarified early on because animal
rights campaigners have had a tendency to react to the 'typical'
biologist by denying that evolution works in the way he claims
it does. I believe that much more progress is possible if we can
agree to accept facts for which evidence exists and to argue
about the way in which these facts should be interpreted.

2

ON THE DIVERSITY OF LIFE

This chapter aims to provide a background to the problem of discussing the nature of animals and their position in human ethics. It will show that some animals are so closely related to us that it is possible that an objective system of taxonomy would classify them as other species of humans, but that there are also some species which are very probably not even sentient. It will demonstrate how inconsistency and over-rigidity in the way we apply systems of classifying animals have led us to permit unnecessary and avoidable suffering.

Any theory of animal rights must accommodate the fact that the word 'animal' covers an immense diversity of tremendously different species of living creatures. There is a good deal of confusion about this, and some of the arguments about the nature and possibility of animal rights are essentially products of this confusion. For example, it is sometimes suggested that the idea of rights for animals is absurd because it would entail granting rights to insects and other invertebrate animals, which is 'obviously' nonsense. This particular conflict seems to stem from a mistake about the nature of definitions of words: if someone who argues for animal rights is actually thinking of vertebrate animals only, then he is guilty of imprecision, but it does not follow that his arguments cannot be good arguments in favour of rights for vertebrates (or mammals, or whatever group he has in mind). It would be useful if everyone talking about animal rights did try to be precise in specifying which particular taxonomic groups were in question. However, an opponent of animal rights cannot validly speak as though he had discovered that the word 'animal' really means any multicellular living organism which is not a plant or fungus; that this includes some animals such as sponges which lack nervous tissue and can hardly be sentient; and that therefore it cannot be true that 'animals' can have rights. It is a fact that common English usage does sometimes use the word 'animal' as a synonym for 'vertebrate', or even for 'mammal'.

In the *Oxford English Dictionary* 'animal' is defined as:

A. 1. A living being; a member of the higher of the two series of organized beings, of which the typical forms are endowed with life, sensation, and voluntary motion, but of which the lowest forms are hardly distinguishable from the lowest vegetable forms by any more certain marks than their evident relationship to other animal forms, and thus to the animal series as a whole rather than to the vegetable series.

2. In common usage: One of the lower animals; a brute, or beast, as distinguished from man. (Often restricted by the uneducated to quadrupeds; and familiarly applied especially to such as are used by man, as a *horse*, *ass*, or *dog*.)

3. Contemptuously or humorously for: A human being who is no better than a brute, or whose animal nature has the ascendancy over his reason; a mere animal.

(Cf. similar use of *creature*.)[1]

Collins Dictionary of the English Language defines an animal as:

1. *Zoology*. any living organism characterized by voluntary movement, the possession of cells with noncellulose cell walls and specialized sense organs enabling rapid response to stimuli, and the ingestion of complex organic substances such as plants and other animals. 2. any mammal, esp. any mammal except man. 3. a brutish person. 4. *Facetious*. a person or thing (esp. in the phrase *no such animal*).[2]

The Cruelty to Animals Act 1876,[3] the Protection of Animals Act 1911, the Protection of Animals (Scotland) Act 1912, and the Welfare of Animals (Northern Ireland) Act 1972 all give definitions of 'domestic animal' and 'captive animal' which

---

[1] *The Compact Edition of the Oxford English Dictionary* (Oxford: Oxford University Press, 1971).

[2] Patrick Hanks (ed.) (London: Collins, 1979).

[3] Now replaced by the Animals (Scientific Procedures) Act 1986, which, instead, defines 'a protected animal' as any living vertebrate other than man. This Act also gives the Secretary of State responsible for its operation the power to extend the definition to particular groups of invertebrates.

specify that the term 'animal' includes bird, fish, or reptile (it is not clear what they make of amphibia), but excludes invertebrates. To an extent, these Acts are mutually inconsistent in their treatment of vertebrates; for example, it was pointed out by the RSPCA's Panel of Enquiry into Shooting and Angling that, if the normal practices involved in angling were carried out in a laboratory on unanaesthetised fish and without a Home Office licence, then this would probably be an offence under the 1876 Cruelty to Animals Act.[4] However, no angler has ever been prosecuted under the 1911 Protection of Animals Act, which deals with cruel treatment of animals outside the laboratory (para. 252). In this connection there is also an inconsistency in the interpretation of the 1911 Act which appears to depend upon the circumstances of the alleged cruelty. The RSPCA obtained a successful prosecution following a theatre play in which a goldfish was deliberately emptied on to the stage to die of suffocation (but was rescued by one of the Society's inspectors), but it appears to be quite legal for anglers to use small fish (sometimes including goldfish) as live bait:

attachment to the tackle . . . involves some degree of impalement. The bait fish is then cast to swim in water where pike are expected until it is taken or until it succumbs to stress or to the injuries inflicted on it in the course of attachment. It is not unusual . . . for a live bait fish to tear free from the tackle, inevitably increasing the severity of its impalement injuries. (Medway Report, para. 161)

A survey carried out by the Universities Federation for Animal Welfare in 1986 revealed that, in the Thames Water Authority area, over half of the angling clubs controlling stretches of water in the area still permitted use of vertebrate live-baiting.[5] (In fact the true numbers may have been higher than this, since 20 per cent of clubs contacted refused to complete the form, and 28 per cent failed to respond to queries. McEwen comments that it is possible that clubs which consider that they have a good record on welfare will be more likely to reply to a questionnáire from an animal welfare organization than

[4] *Report of the Panel of Enquiry into Shooting and Angling (1976–1979)*, chairman, Lord Medway (Horsham: Panel of Enquiry into Shooting and Angling, 1980).
[5] Peter McEwen, *Vertebrate Livebaiting* (Potters Bar: Universities Federation for Animal Welfare, 1986).

those which feel they could be laying themselves open to criticism.) Treating birds or land mammals like fish would be regarded as obvious cruelty.[6]

In British statute law the definition of 'animal' is sometimes clearly being restricted to permit concise wording, rather than because those drafting the legislation actually considered that 'animal' has this meaning in normal usage. For example, the Diseases of Animals (Milk Treatment) Order No. 1714 (1967), defines 'animal' as all ruminating animals and swine, presumably to avoid inconvenient repetition. At other times (for example the 1911 Act) it does seem that invertebrates are genuinely not seen as 'animals' at all, and the legislators seem to be puzzled over the position of birds (for example, the Wildlife and Countryside Act 1981 deals with birds and 'animals' (animals which are not birds) as two classes).[7] Such legislative redefinition of 'animal' opens the possibility of a similar redefinition for the purpose of legal or moral animal rights (for example, as vertebrates plus some specific groups of invertebrates) to fall in line with current knowledge about the likelihood of sentience in the various classes of living things.

In discussing how we might make decisions about the kinds of living things which are probable candidates for rights-bearing status I shall make frequent reference to various kinds of scientific evidence, some of which may have involved suffering for the research subjects. It has been suggested that philosophers who criticize some kinds of research on animals cannot reasonably make use of such research to provide a factual basis for their arguments in favour of the sentience of animals.[8] However, I think we would not suppose it was unreasonable to say that some kinds of evidence which might be provided by scientific research could lead us to change our minds about the legitimacy of subsequent research. (For example, we might start off believing that human and animal

[6] Although until quite recently it was normal practice for British whalers to inflict very similar injuries on their quarry, and modern whaling nations still catch minke whales by the 'cold' (non-explosive) harpoon, which causes a slow death from exhaustion and loss of blood following impalement.

[7] Wendy Crofts, *A Summary of the Statute Law Relating to Animal Welfare in England and Wales* (Potters Bar: Universities Federation for Animal Welfare, 1984).

[8] Personal comment to the author by Dr J. Stewart.

brains and sensory systems must be quite different, and scientific evidence that they are sometimes very similar could make us subsequently feel that procedures which would cause pain to a human would be likely to cause equivalent pain to an animal.) Conversely, my use of evidence obtained from painful experiments does not necessarily imply that I think that these were justifiable in the first place.

Biologists themselves are not in complete agreement about the kinds of creature which should be classified as animals: some would place both sponges and protozoa in the Kingdom Animalia, while some consider that these two groups should each have their own separate kingdom. In order to explain the nature of these disagreements it is first necessary to say a little about biological nomenclature. Scientific names are Latin in form, written in the Roman alphabet and according to the rules of Latin grammar, although they may be derived from non-Roman languages (for example, the scientific name of the gorilla is *Gorilla gorilla*, both parts of the binomial being derived from the local African name for this ape). A particular species is uniquely labelled by a pair of terms: the name of the genus (group of closely related organisms) to which the species belongs, followed by a species name. The species name cannot be used alone, since (as, for example, in the many kinds of plants with the second term 'japonica') it may be shared by several species. Once the full Latin name of a species has been cited in a particular text it is standard practice to abbreviate the generic name to its initial letter (for example, *Pan troglodytes*, the common chimpanzee, would be abbreviated to *P. troglodytes*). Organisms are classified in a hierarchical system, in which groups become more and more inclusive as one passes along the series from variety to species to genus, and so on, up to the widely embracing kingdom, whose members share only their most general characteristics.

The complete range of levels will not always be used in any particular discussion of a problem of classification.[9] Some categories have standardized endings, for example, in zoological classification, superfamily names end -oidea, family names -idae, subfamily -inae, and tribe -ini; for example: Hominoidea—the

[9] The taxonomic categories are kingdom, phylum, class order, family, tribe, genus, species.

superfamily of apes and man; Hominidae—the family of 'men'; Lorisinae—the subfamily of prosimians which contains the potto and slow loris. Unfortunately, botanists, zoologists, and bacteriologists do not necessarily use identical terminology, and difficulty is further compounded where an organism's status as a plant or animal is in dispute (as, for example, the photosynthetic but motile single-celled euglenids, which might be classified as algae or animals, depending on the particular persuasion of the biologist concerned.[10]

Part of the disagreement about defining what an animal is stems from conflicting interpretations of the evidence about relationships between different groups of living organisms: some biologists consider that sponges probably developed the multicellular habit independently of other metazoans (multicellular animals); while others think that the earliest multicellular animals would have been rather similar to sponges. Sponges appear in many ways to be intermediate between true metazoans and colonial protozoans, like the photosynethetic *Volvox*. They lack organs, defined tissue layers, digestive sacs, nerves, muscles, and sense organs.[11]

More fundamentally, there is no general agreement on whether classificatory terms like 'animal' should represent evolutionary relationships, or whether they should indicate the possession of similar characteristics (such as motility, heterotrophism (use of complex chemicals for nutrition), or sensitivity). Groups of organisms which have important similarities but probably do not have a common evolutionary origin are termed polyphyletic. Groups which do have a single common evolutionary origin are termed monophyletic. Because the genetic code is universal for all known terrestrial organisms, it is thought that all living organisms are fundamentally monophyletic in origin, i.e. the events which produced terrestrial life happened only once. However, it is not known whether the organisms which biologists commonly place within the Kingdom Animalia are monophyletic or polyphyletic. The sponges (Poriferans), the Mesozoa (tiny multicellular parasites), the

[10] Charles Jeffrey, *Biological Nomenclature* (London: Edward Arnold and The Systematics Association, 1977).

[11] Robert H. Barth and Robert E. Broshears, 'The Sponges', in *The Invertebrate World* (New York: CBS College Publishing, 1982), 71–85.

Placozoa (amoeboid organisms composed of a flattened disc of about 1,000 cells), and the coelenterates (jellyfish and sea anemones) have been suggested as groups which may have arisen from protozoan ancestors, independently of the line derived from flatworm-like precursors to which we belong.[12] Members of the phyla Porifera, Placozoa, and Mesozoa lack both nervous and special sensory systems.[13] So far only one species of the phylum Placozoa has been identified. Each organism is composed of a small (3 millimetre diameter maximum), flattened, hollow ball of cells filled with fluid in which some cells float free. There is no nervous system, but the organism is able to crawl like an amoeba by changes in body shape. Members of the phylum Porifera are sessile (rooted) organisms, lacking any nervous system and feeding by filtration of particles in water drawn through pores of the sponge by the beating of the flagella of choanocyte cells. The latter are very similar to the free-living single-celled organisms known as choanoflagellates, and it is surmised that sponges may have evolved from such organisms by development of a colonial life-style.

There are two coelenterate phyla: Cnidaria and Ctenophora. The Cnidaria are free-swimming or sessile organisms which possess a nervous system, although this is not concentrated into a central processing system of any kind, being rather a diffuse network of linked cells running through the body of the organism. The Ctenophora (comb jellies) are free-swimming, predatory organisms, basically similar to the free-swimming forms of the Cnidaria in their nervous organization. Mesozoa are small, internal parasites, worm-like in shape, but lacking any muscular or nervous system.

Viruses are non-cellular, being composed essentially of a string of nucleic acid (either DNA or RNA) enclosed in a protective coat of protein. They are very much smaller than bacteria, and in some cases consist merely of a unit of nucleic acid. In some cases, the viral genome (total genetic information)

[12] R. S. K. Barnes, 'Kingdom Animalia', in R. S. K. Barnes (ed.), *A Synoptic Classification of Living Organisms* (Oxford: Blackwell Scientific Publications, 1984), p. 131.
[13] Barnes, *A Synoptic Classification of Living Organisms*, pp. 133, 134, 158. Barth and Broshears, *The Invertebrate World*, pp. 69–84.

may even be divided between several particles, which must enter a cell together if infection is to occur. Viruses only contain one type of nucleic acid and must rely on the host cell for the 'read-out' of the information contained to produce new viral material.[14]

Note, also, that slime moulds, which are nominally classed as Fungi, show the characteristics of motility and co-ordination to at least as great an extent as mesozoans and placozoans; while some vascular plants, such as the Venus fly-trap and the sensitive plant *Mimosa pudica*, seem to approach the complexity of some coelenterates.

It is evident that if we believe that the capacity for sensation is an important factor in deciding whether an organism has moral status we will concentrate our attention upon those groups which possess significant nervous development. (But we need the biological evidence to identify the relevant groups.) Evolutionary relationships are significant from a biological viewpoint, but when considering the moral status of particular kinds of beings their existing characteristics will be more important than historical origins. A creature which had evolved on another planet would not share any evolutionary continuity with terrestrial life; hence in terms of phylogeny it could not be plant, animal, or bacterium. In terms of functional characteristics, however, we should most probably want to place it within these broad groupings in a purely analogous capacity. Hence a non-sentient, but multicellular and auto-trophic organism would be broadly classed as a type of plant; multicellular, sentient, motile organisms would be analogous to terrestrial animals; and so on. In terms of rights and duties we are primarily concerned with a particular being's mental characteristics, irrespective of its history. It seems that only sentient beings are plausible candidates for rights.[15] Only

---

[14] J. L. Cooper and F. O. MacCallum, *Viruses and the Environment* (London: Chapman and Hall, 1984), 1–24.

[15] C. S. Lewis notes on this question, 'The fact that vegetable lives "prey upon" one another . . . in a state of "ruthless" competition is of no moral importance at all. "Life" in the biological sense has nothing to do with good and evil until sentience appears. . . . A forest in which half the trees are killing the other half may be a perfectly "good" forest: for its goodness consists in its utility and beauty and it does not feel' (*The Problem of Pain* (Glasgow: Collins, 1957), 118).

sentient creatures have the characteristic ability to experience value which seems to be so crucial in deciding what entities have moral significance.

Because of these problems of definition, unthinking use of the term 'animal rights' immediately involves us in a degree of muddle, confusion, and imprecision. I should prefer to say instead that we are beginning to realize (partly perhaps due to horizons opened by the imminence of human exploration of space)[16] that there exist other organic minds besides human ones, and that these deserve some kind of moral consideration on our part. Since the only currently plausible candidates for non-human minds in our immediate vicinity happen to be non-human animals, this consideration then involves us in thinking about the rights of animals, but it is quite important to recognize that there is nothing intrinsically special about *animal* rights versus (for example) the rights which would belong to a conscious plant, sentient electronic computer, or Martian; those which we should apply to human foetuses which have reached a state of development at which it is probable that they are capable of sensation;[17] or even quite bizarre possible entities, such as schizophrenic secondary personalities. At first sight, non-human animals appear very different from the pictures we have of intelligent aliens. I think this difference is due to our preconceptions about what a non-human mind should be like, plus a tendency for familiarity with animal minds to breed contempt. Alien intelligence is normally (I think) expected to be either equal to, or greater than, human intelligence, but there is no real reason why we should put human average intelligence as the 'base level' for

[16] See e.g. essays by Michael Ruse, 'Is Rape Wrong on Andromeda? An Introduction to Extraterrestrial Evolution, Science and Morality', and Jan Narveson, 'Martians and Morals: How to Treat an Alien', in Edward Regis, Jr. (ed.), *Extraterrestrials: Science and Alien Intelligence* (Cambridge: Cambridge University Press, 1985), 43–78, 245–65. Both essays make the point that the problem of how to treat sentient extraterrestrials is essentially identical to that of the proper attitude to sentient non-human terrestrials (unless the extra-terrestrials are cleverer than we are, in which case the matter is likely to be their moral problem, not ours).

[17] The Animals (Scientific Procedures) Act 1986 provides for statutory protection of mammalian foetuses after half the gestation period has elasped, presumably on the grounds that this marks the stage at which it becomes likely that the foetus is capable of sensation. Abortions of human foetuses in the latter part of the legal period fall within this time limit.

qualification as a rational being. No doubt there were species ancestral to humans who had an average intellect somewhat below that of present-day *Homo sapiens* but whom we should wish to classify as people not animals. The type specimen of *Australopithecus africanus* is referred to in palaeontological literature as the Taung child, although his/her brain-case is no larger than that of a young chimpanzee.[18]

With these caveats, in this work I intend to adopt the common biological definition for 'animal' (multicellular, not a plant, nor a human). This definition excludes bacteria and viruses, which are sometimes considered as animals (though not normally so by biologists), because they lack the organized cell structure of nucleus and cytoplasm seen in plants, protozoa, and metazoans, and are actually less like animals than are multicellular plants.[19] It also excludes protozoa (single-celled organisms with organized eukaryote cell structure), which some biologists would group with animals, on the grounds that multicellular structure seems to be essential for the possibility of some properties we associate with animals which cause us to consider them as possible rights-bearers. An amoeba is essentially similar to a free-living version of a human white blood cell; the association of many cells to form multi-cellular bodies seems to me to be essential for the possibility of sentience.

Within the group of animals biologists normally recognize a fundamental distinction between vertebrates and invertebrates. The latter are generally smaller, less complex, and less capable of 'intelligent' learned behaviour than vertebrates, although there are important exceptions, like the cephalopods, which are relatively large and exhibit learning capacities within the range of that shown by small mammals such as mice and rats.

Comparisons between different groups are difficult because some small animals have special adaptations which allow efficient functioning of relatively tiny nervous systems: for example, some invertebrates achieve smooth control of muscle tension by altering the rate of firing of single nerves, where

[18] Yoel Rak, *The Australopithecine Face* (New York: Academic Press, 1983).

[19] Plants, protozoa, and animals are commonly classified as eukaryotes, while bacteria are classified as prokaryotes.

vertebrates have many nerve fibres running to each muscle and control tension by varying the number of active fibres.[20] Vertebrates may have minute brains, like some of the smallest fish, or huge ones, like humans, elephants, and whales. Some, like the chimpanzee and gorilla, have been distinct from the human line for no more than a matter of a few million years. It is commonly accepted that the chimpanzee–gorilla–human split occurred around 8–10 million years ago.[21] Human, gorilla, and chimpanzee lines have been separate for approximately the same length of time as those of brown, black, and sun bears (*Ursus arctos*, *Ursus americanus*, *Helarctos malayanus*); and the African ape–human line diverged from that of the orang-utan at about the time of the split between the ursid line and the genus *Tremarctos* (spectacled bear).

Tom Regan attempts to eliminate the problem of coping with different kinds of animal life by concentrating on arguments in favour of the rights of mammals one or more years old (as creatures we are reasonably justified in supposing to be 'subjects of a life').[22] However, this produces the odd result of committing him to the idea that some small mammals (such as mice and rats) become the possessors of moral rights when they are middle-aged, and it eliminates birds from consideration altogether, in spite of the growing evidence that birds and mammals have very similar intellectual capacities.[23] Some bird species have been shown to be capable of tool use; many can solve simple reasoning problems; and an African Grey parrot has demonstrated a capacity to learn flexible use of a simplified language system which rivals that of chimpanzees.[24]

[20] M. J. Wells, *Lower Animals* (London: Weidenfeld & Nicolson, 1968), 53.

[21] S. J. O'Brien, W. G. Nash, M. E. Wildt, M. E. Bush, and R. E. Benveniste, 'A Molecular Solution to the Riddle of the Giant Panda's Phylogeny', *Nature*, 317 (12 Sept. 1985), 140–4.

[22] *The Case for Animal Rights* (London: Routledge & Kegan Paul, 1983), 77–81.

[23] For a brief, semi-popular review of the evidence for this see Joe Crocker's article, 'Respect Your Feathered Friends', *New Scientist*, 1477 (10 Oct. 1985), 47–50.

[24] Irene M. Pepperberg, 'Acquisition of Anomalous Communicatory Systems: Implications for Studies on Interspecies Communication', in Ronald J. Schusterman, Jeanette A. Thomas, and Forrest G. Wood (eds.), *Dolphin Cognition and Behavior: A Comparative Approach* (Hillsdale, NJ: Lawrence Erlbaum Associates, 1986), 289–302.

Traditionally, Indian philosophical systems have devoted more time to consideration of the status of animals than those of Europe and America. It is interesting, therefore, that some of these systems classify living things according to their supposed degree of sentience and rate the wrongness of harms on a sliding scale according to the sentience of the victim. Thus mammals such as men, horses, and cows are considered to be more sentient than insects, which in turn are more sentient than plant life. Killing a member of the first class is murder; killing plants is not sinful at all; while harms to entities in between are more or less evil according to their place on the scale.[25] Westerners perhaps tend to assume that Indian views on killing animals are primarily religious in character: the superstitious observance of pointless rules, or a manifestation of profound spirituality, depending on how sympathetic the individual is to Indian culture. I'm not sure that this is necessarily true, or at least that Indian views on animal life are more importantly grounded in specifically religious belief than Western views about human life. We do frequently voice our belief in the wrongness of (for example) infanticide in religious terms, 'the *sanctity* of human life', but I doubt whether we would normally want to say that this belief is primarily religious or mystical in character. More importantly it relates to our views about the nature of human beings, although it is true that these views are at least partly shaped by the philosophical picture of social relations which is inherent in our Christian-based culture. Judaeo-Christian culture tends to emphasize how different humans are from animals while Hindu, Jain, and Buddhist traditions emphasize how similar they are. It seems to me that these contrasting views about the nature of animals will naturally tend to produce contrasting views about how they should be treated. We have to have some theory about the kinds of entities we deal with and such theories have tended to be based on religious teaching. However, the different ethical consequences derived from contrasting theories will be religious only in a weak sense.[26] Lodrick cities the nineteenth-century

---

[25] D. O. Lodrick, *Sacred Cows, Sacred Places: Origins and Survivals of Animal Homes in India* (Berkeley and London: University of California Press, 1981), 154–5.

[26] Someone who believes that abortion is wrong may well say that God will

French author Rousselet, who visited a *pinjrapole* (animal hospital) in Gujarat: 'Some of these animals appear to be so sick that I venture to tell my guide it would be more charitable to put an end to their suffering. "But", he replies, "is that how you treat your invalids?".'[27] This is esssentially a secular response rather than a religious one.

A few 'case-studies' of selected species will illustrate the range of types which are included in the broad classification of 'animals' and the impossibility of producing any simple, all-embracing theory which will be applicable to any animal irrespective of species.

Aplysia is a genus of marine slugs, not too distantly related to the common pulmonate terrestrial slugs. From both biological and philosophical viewpoints, these relatively simple animals are interesting because they are the only creatures with organized 'brains' for which we possess something approaching a complete 'wiring diagram'.[28] We know that aplysians behave in a physically determinate way, which can be entirely explained in terms of the mechanical activity of their nervous systems. The aplysian central nervous system is composed of nerve-cells which operate in basically the same way as those of vertebrates, but are arranged rather differently. Instead of one central brain, the cells are grouped into clusters (ganglia), linked by connective nerve fibres. There are two main groups of ganglia: the head ganglia, which form a ring surrounding the animal's oesophagus, and the visceral ganglia, which are situated further down in the body tissues. The former largely control 'voluntary' activities, such as feeding, swimming, and movement of the head, and also sense reception. The latter mainly control hormonal processes, such as regulation of the animal's water balance, egg-laying, and inking.[29] Thus at one level (cell structure) the aplysian brain is very like our own, while at another (cell arrangement), it is considerably different

punish people who perform abortions, but I suspect that he would say also that God *justly* punishes abortion, rather than that abortion is wrong because God disapproves.

[27] Lodrick, *Sacred Cows, Sacred Places*, p. 69.

[28] Eric R. Kandel, *Behavioral Biology of Aplysia* (San Francisco: Freeman, 1979).

[29] Ibid., pp. 119–38

in basic organization, as well as being very much smaller. On the other hand, comparative psychologists still find it useful to describe the slug's behaviour in terms of motivational states[30] (such as hunger and satiation), and we have no way of knowing whether the animals experience any of the subjective states we normally associate with these terms.[31]

Since we have no knowledge of the way in which physical brains can generate mental experiences, our relatively great knowledge of the mechanics of *Aplysia*'s brain does not help us very much with the question of its ability to feel pain or pleasure. Possibly, the most we can say is that, since we know that a reduction in usable brain tissue decreases the range of possible experience in humans, the experience of *Aplysia* must at all events be far simpler than our own. Furthermore, we can be certain that *Aplysia* does not have 'free will'. However, since a physicalist account of human experience must assume that we do not have it either, this is perhaps of no very great significance.

Octopus vulgaris (common octopus) is another mollusc, a member of the group Cephalopoda (literally head-foot); they are active animals, whose central nervous systems are built on a basic plan of nerve-cell aggregations surrounding the gut, like their simpler relative *Aplysia*, but much enlarged and concentrated, so that they approach the complexity and connectivity seen in some vertebrates. The brain: body weight ratio of *Octopus* is at least equal to the average for fish and reptiles; most squids have even higher values, and, in any case, since the cephalopod brain is mainly concerned with high-level sense integration and with movement control, other activities being delegated to lower ganglia, the amount of brain available to a cephalopod for learning may equal or exceed that of some birds and mammals.[32] *Octopus* has considerable powers of learning and sensory discrimination, although with some odd gaps from a vertebrate point of view; for example, the animals

[30] Kandel, *Behavioral Biology of Aplysia*, p. 358.

[31] See Kandel, *Behavioral Biology of Aplysia*, pp. 358, 367, 369, for proposed wiring diagrams of the neural control of sensitization (generalized arousal of the animal in response to noxious stimuli), feeding, and defensive behaviour in *Aplysia* and *Pleurobranchaea*, another marine slug.

[32] M. J. Wells, *Octopus* (London: Chapman and Hall, 1978), 7–8.

appear to find it very difficult to learn particular postural movements and to discriminate between objects which differ only in their weight, possibly because the absence of a rigid skeleton makes the number of possible attitudes too great for efficient monitoring.[33] Discrimination between different weights requires the ability to monitor precisely the relationship between body position and muscular effort expended when lifting an object. These factors mean that *Octopus* is very limited in ability to carry out manipulative tasks in spite of the promising combination of flexible arms and an intelligent brain. Lever-pulling tasks which rats readily accomplish are performed poorly, although M. J. Wells cites a rather engaging account by P. M. Dews of an octopus ('Charlie') who solved the task of pulling a lever to activate a light under which he was then fed by attempting to pull the light into his tank, squirting the experimenter with water, and finally dismantling the lever apparatus. In nature, octopuses will draw stones behind them upon retreating into their holes, making a simple shield, but never appear to show any genuine building activity.[34]

Similarly, another member of the group, the cuttlefish (*Sepia officinalis*), which normally catches prey using two long tentacles, does not seem able to 'see' that grasping fast-moving objects is no longer possible if these tentacles have been severed at the base. The animal continues to make abortive strikes at prawns, although it is quite possible for *Sepia* to learn to inhibit attacking behaviour when this is punished by electric shock.[35] This example of inability to recognize the loss of a limb is perhaps not as unlike vertebrate behaviour as it might appear. Humans who have undergone amputation sometimes experience 'phantom limbs', and baby mice who have had their forelimbs amputated at birth continue to make ineffective grooming motions with the stumps at the time when they would naturally develop the ability to groom themselves, although they cannot reach their faces.[36]

---

[33] Ibid., pp. 243–5.   [34] Ibid., p. 241.

[35] J. B. Messenger, 'Prey-Capture and Learning in *Sepia*', in Marion Nixon and J. B. Messenger (eds.), *The Biology of Cephalopods* (London: Academic Press, 1977), 358–63.

[36] The latter experiment was performed in America, and it is most unlikely that it would have been permitted under British law. Since cephalopods are invertebrates, British law did not recognize them as animals for the purposes of

Because of the fundamental structural differences between
the brains of cephalopods and vertebrates it is not possible to
relate any distinct brain regions in these animals to the areas
which are known to be involved in sensation and emotion in
humans. The human brain is known to have the same basic
plan as all other vertebrates, and differs only in the enormous
expansion of cell numbers. Cephalopods have developed complex nervous systems from simple ganglia quite independently
of the vertebrate line. The basic neuronal building blocks are
the same in both cases, but their arrangement has been dictated
by differing circumstances, although there are some clear indications of functional convergence. One example is the eye, which
has arisen independently in the two lines: the vertebrate retina
is back to front in comparison with that of cephalopods, presumably because of differences in the primitive structures from
which the two systems evolved. However, in spite of this difference, both groups contain members with functional binocular
vision and depth perception as judged from their behaviour[37]
and this involves the integration of information from the two
sides of the brain.[38] Because of this, the question of consciousness and its possible qualities is more problematic for this
group of animals than for our fellow vertebrates. Some of the
senses we know that *Octopus* possesses are impossible for us
to imagine: for example, what can it be like to be able to 'taste'
with one's body surface? What is it like to be less, not more,
ready to seek food when deprived of it for some time?[39] In spite
of this, some researchers have found it useful to characterize
octopus behaviour in terms of motivation and emotion (although
it is unclear whether they intend this in a purely behavioural
sense, rather than in terms of subjective experience on the part
of the animal). In his paper on the effect of fear motivation on
learning in *Octopus*,[40] G. D. Sanders describes an attempt to

the Cruelty to Animals Act 1876, or the Protection of Animals Act 1911. (The
present Animals (Scientific Procedures) Act still does not include the
Cephalopoda as protected species under the Act.) Hence the use of cephalopods
for scientific purposes is not subject to legal regulation in this country.

[37] Messenger, 'Prey-Capture and Learning in *Sepia*', pp. 349–58.

[38] Ibid., p. 357.

[39] Wells, *Octopus*, pp. 8–9.

[40] G. D. Sanders, 'Multiphasic Retention Performance Curves: Fear or
Memory?, in Nixon and Messenger (eds.), *The Biology of Cephalopods*,
pp. 435–44.

investigate the separate effects of forgetting and fear on the animals' ability to remember the solution to a simple test. The octopus was required to learn to avoid picking up a wooden ball which was marked with carved grooves. Wrong behaviour was punished by electric shocks. Picking up a smooth ball did not result in punishment, and the animals learned to discriminate between the two types of ball. After training, the animals' memory of the discrimination between the two objects was tested at various times and the researchers were able to show that (like rats) their performance declined dramatically over a period of one to six hours, but then improved once more. In rats this effect is considered to be caused by a combination of two factors. Firstly, it is thought that the learned discrimination is initially stored in short-term memory and only transferred into long-term memory after several hours have elapsed. Animals perform poorly at intermediate times because their short-term memory has begun to deteriorate, while the long-term trace is not yet fully established. Secondly, because training by electric shock causes fear, part of the apparent learning is due to a non-specific increased timidity which inhibits any kind of active movement, including performing the punished act. It is suggested that during the hours shortly after initial training there is a rebound effect (parasympathetic over-reaction) which abnormally *decreases* fear. This could act on both the trained fear of the punished action and the non-specific fear, and both effects would have a tendency to produce the observed increase in performance of the punished behaviour. Sanders attempted to analyse these effects in *Octopus* by testing groups of animals at 0.5, 2, and 4 hours after training using either a rough ball (acceptance punished) or a smooth one (acceptance rewarded with a piece of fish). The 2-hour group had the greatest number of errors (takes of the rough ball), but all groups showed an equal decrease in takes of the smooth ball relative to animals who had not been punished. Sanders argues that this indicates a genuine memory deficit at two hours following training, since the equal decrease in takes of the smooth ball indicates that all three groups had similar levels of non-specific fear. In view of the conclusion that *Octopus* experiences non-specific fear when given electric shocks one might perhaps question the justification for continuing similar experiments, particularly since (in order to eliminate visual-learning effects) the animals were blinded beforehand (under anaesthetic). If cephalopods were classified as protected animals for the purpose of the regulatory legislation, one suspects that it would be obligatory for researchers to eliminate visual clues by darkening the aquaria in which tests were done rather than by eliminating the animals' sight permanently. However difficult it may be to know exactly what an octopus feels after being blinded, Wells's description portrays something with which it is possible for a human to empathize:

generally it is necessary to keep the animals for several days after blinding before the beginning of training. During this time they at first sit huddled up with their arms curled tightly around them. If disturbed they may roll over to present the suckered underside, as an octopus does if poked in its hole in the sea. Within a day or so the blinded animals begin to feed regularly, accepting small pieces of fish or crabs touched against the arms. Within a week they will normally be found sitting on the walls or floor of their tanks, with the arms outstretched. They now move towards any source of disturbance and will once again grasp and pass small objects towards the mouth.[41]

Ovis aries. Sheep, in a sense, can be said to be representative of the 'typical' kind of animal which forms the subject of most discussions of animal rights. As social mammals who have played a substantial role in human life for over 11,000 years,[42] they are at once easier for us to identify with than 'alien' forms like the invertebrates, and sufficiently distanced that they do not generate the admixture of hostility shown towards the apes. Sheep form strong bonds towards one another, and exhibit distress when separated. If attachment to other sheep is prevented by forcible separation, other animals can become substitute sources of companionship. This cross-specific bonding does not depend upon rewarding by food (as in suckling by the mother animal), since young lambs can be induced to become attached to a dog if the two are confined together. (They will

---

[41] *Octopus*, p. 220.

[42] See M. L. Ryder's mammoth work *Sheep and Man* (London: Duckworth, 1983) for a detailed account of the origin of sheep and their part in ancient and modern societies throughout the world.

follow the dog, become disturbed when separated, and so on.)[43] The opportunity to resume contact with the dog can then act as a reward in training to reach the goal area of a simple maze.

Valerius Geist found that the wild American bighorn sheep in his study group had excellent memories. Sheep who had become accustomed to come to him for salt would approach willingly even after a gap of several months without contact. When he hid from sight the sheep did not search the area where he had disappeared, but instead attempted to intercept him at the place where he should have been if he had continued on a straight line: a feat which requires insight rather than mere association.[44]

In pre-industrial societies sheep were frequently the major source of raw materials—blood, milk, wool, dung, meat, fat, bone, skin, horn, and gut—as well as being used as pack animals on occasion.[45] In general, the enforced association with human society has not involved a very great restriction upon the animals' natural social behaviour: wild sheep normally live in separate male and female herds, which associate during the breeding season, and these flocking tendencies have generally been allowed to continue under human control, with few attempts to create more intensive systems of management as with poultry and pigs. Hulet and Hafez concluded that the only significant change in the social organization of domesticated sheep relative to their wild relatives is that domestic flocks are generally larger, and somewhat disorganized, due to human admixture of unrelated animals, with the effect that related animals are frequently separated, causing continual bleating as they attempt to regain contact—something which would be likely to attract predatory attentions in the wild state.[46] The major change in the life of sheep which has been caused by human selective action lies in the development of the typical woolly fleece from a short, hairy, wild-type coat, not very different from that of goats or cattle. This abnormality does

---

[43] C. V. Hulet, G. Alexander, and E. S. E. Hafez, 'The Behaviour of Sheep', in E. S. E. Hafez (ed.), *The Behaviour of Domestic Animals*, 3rd edn. (London: Ballière Tindall, 1975), 280–2.

[44] *Mountain Sheep* (Chicago and London: University of Chicago Press, 1971), 43.

[45] Ryder, *Sheep and Man*, pp. 712–13.

[46] 'The Behaviour of Sheep', pp. 246–94.

restrict the animals' ability to control their body temperature: modern breeds of sheep do not moult in the summer to any extent; wool cannot be flattened or fluffed up like normal hair to reduce or increase heat conservation, although some temperature measurements suggest that a thick coat of wool can act as an efficient insulator against heat from sunlight, and that domestic sheep have adapted to regulate heat loss and gain in new ways. Having wool instead of hair may make young lambs more likely to suffer from chilling immediately after birth.[47]

Some critics of human use of animals[48] have claimed that the creation of this kind of abnormality is such a serious encroachment upon the animals' capacity to lead a satisfactory life that we ought to cease to breed from such animals and allow them to become extinct. The same argument has also been applied to most other domestic animals; particularly to some breeds of dog (such as the Pekinese) which exhibit physical traits which would be considered pathological in members of the wild ancestral species.

Elephants (*Loxodonta* and *Elephas*). Most mammals' brains are around 90 per cent of the adult weight at birth. However, in elephants the figure is about 35 per cent (in man the corresponding figure is 26 per cent), and this seems to be responsible for the considerable learning ability shown by young elephants. The adult mass is greater than that of any other land mammal (including man), although obviously the brain : body ratio of a 3,000 kg elephant with a 4 kg brain will be less than that of a 100 kg man with a 1.5 kg brain. 'Average' brain : body ratio falls, however, with increase in body mass: when different mammalian species' values are plotted graphically the elephant falls above the expected value for an 'average' mammal (for comparison, the average body-weight of a male African elephant is about 73 per cent more than that of a female but the brain-weight difference is not much more than 20 per cent of the female value).[49]

---

[47] Geist, *Mountain Sheep*, p. 243.

[48] e.g. John Bryant, *Fettered Kingdoms* (Privately printed, 1980), and John Rodman, 'The Liberation of Nature', *Inquiry*, 20 (1977), 83–145.

[49] Elephant brain- and body-weight values taken from S. K. Eltringham, *Elephants* (Poole: Blandford Press, 1982), 16–17, 127–8.

Elephant society is fundamentally matrilineal and the old female has a status far higher than is usual in most animal species. The unit of society is the family, which consists of a number of mature females and their calves. The adults are closely related to each other, being either mother and daughters, or sisters, half-sisters and cousins. Sometimes one old female, the matriarch, is very much older than the rest. She is probably the mother, or grandmother, of the others and she is very much the leader. In times of trouble she will defend the group and the rest look to her for guidance. If she runs away, they will follow her but, if she turns round and attacks, they will charge with her. This was often brought home very forcibly to me during immobilization programmes. If one darts a young female in the group, it is extremely difficult to chase off the matriarch. She will stand guard over the stricken animal, trumpeting loudly and attempting to lift it to its feet. The others follow her example, and one is then faced with a milling mass of furious beasts. One tries to avoid this debacle by waiting for a female to wander off a little way before darting her and hoping that the matriarch will not have noticed. An alternative is to dart the matriarch herself. The difference in the reaction of the elephants is striking. In this situation, they seem to have no idea of what to do and rarely make show of resistance as they are shooed away to stand in a forlorn group, anxiously watching the proceedings. Their first reaction when the matriarch falls is to run to her, and this is put to use in culling operations in which whole family groups are usually taken out.[50]

Eltringham also notes that it is the matriarch who stores knowledge of dry-season feeding grounds and of water holes, and that elephant survival depends upon the traditions which she 'hands down' to her group.[51] (He considers that no other animal besides man is so dependent upon traditional knowledge, and that, for this reason, ivory poaching in which the large-tusked matriarchs are preferentially taken out poses an immensely serious threat to the species.)

It seems evident that animals like sheep and elephants, for whom social bonds are such an important part of life, can be harmed in ways which do not apply to solitary creatures like *Octopus*. Killing a social animal not only deprives that individual of future enjoyment of life, but has effects upon all the other members of the group, who may suffer physical or mental distress.

[50] Eltringham, *Elephants*, pp. 52–3.   [51] *Elephants*, pp. 53–5.

The great apes. There are four extant species of great ape: chimpanzee (*Pan troglodytes*); pygmy chimpanzee (*Pan paniscus*); gorilla (*Gorilla gorilla*); and orang-utan (*Pongo pygmaeus*). Morphological, biochemical, palaeontological, and behavioural evidence tends to group these apes into three divisions— African apes: chimpanzee, pygmy chimpanzee, and gorilla; Asian great apes: orang-utan; and the lesser apes: gibbons and siamang. Human and chimpanzee DNA sequences have diverged by only 2.5 per cent since their last common evolutionary ancestor.[52] Human and gibbon DNA differs by 6 per cent; human and rhesus monkey by 10 per cent; and human and squirrel monkey (a New World monkey) by 15 per cent.[53] Chimpanzees' summed genetic difference from humans, measured by comparing the proteins coded by genes at forty-four loci was only 25–60 times greater than the difference between human races.[54] Gribbin and Cherfas make the interesting suggestion that the chimpanzee and gorilla could be descended from australopithecines (putative human ancestors) which became re-adapted to forest life while the proto-human lines remained in the harsher savannah.[55] They further suggest that this could explain why chimpanzees seem to possess human-like reasoning and linguistic abilities, which they develop only under experimental human tuition, and are not seen fully developed in the wild.[56]

Comparison of the molecular sequence of various proteins, such as the haemoglobins, indicates that the African apes are

[52] D. E. Kohne, 'Evolution of Higher Organism DNA', *Qutarterly Review of Biophysics*, 3 (1970), 327.

[53] J. Gribbin and J. Cherfas, *The Monkey Puzzle* (London: Bodley Head, 1982), 114–15.

[54] E. O. Wilson, *On Human Nature* (Cambridge, Mass. and London: Harvard University Press, 1978), 25.

[55] *The Monkey Puzzle*, pp. 181–5.

[56] Ibid., pp. 229–30. But see Yoel Rak, *The Australopithecine Face*, who presents evidence from comparison of the facial skeleton of the australopithecines, apes, and *Homo* which suggests that *A. robustus* and *A. africanus* (proposed by Gribbin and Cherfas as ancestors of the gorilla and chimpanzee respectively) are highly specialized for chewing low-grade vegetable matter, while both African apes and *Homo* retain more generalized primate features and are likely to be derived from a less specialized ancestor, possibly similar to the older *A. afarensis*, which is very similar to modern chimpanzees in facial structure. Rak does not speculate whether *A. afarensis* is ancestral to both *Homo* and the apes, or whether *afarensis* gave rise to *Homo* with the apes diverging previously.

more closely related to *Homo sapiens* than they are to either the orang-utan or the gibbons.[57] It is arguable that an objective classification system would place *Gorilla*, *Pan*, and *Homo* together in the subfamily Homininae; *Pongo* as the only extant genus in the subfamily Ponginae; and the two subfamilies together in the family Hominidae (at present *Gorilla*, *Pan*, and *Pongo* are lumped together in the family Pongidae while *Homo* is given a separate family, the Hominidae).[58]

On this basis, one would have to say (by analogy with similar groupings of genera of, say, bears, or elephants) that there is not one living species of 'man' which may be contrasted with 'animal' species, but four (*Homo*, two species of *Pan*, and *Gorilla*), or perhaps five if we consider that the orang's membership of the Hominidae is a sufficient qualification. Dene *et al.* note that the molecular differences between members of the African ape–human grouping are comparable to those seen within a present-day subfamily grouping of old-world monkeys.[59]

These suggestions raise the natural question: If we come to see the other great-ape species as men does this change our feelings about their status as potential rights-holders? What would be our views if some phylogenically closer relation like *Homo habilis* had also survived into modern times?

It is perhaps interesting to compare Rousseau's picture of human society with the behaviour of chimpanzees, who also appear to form pacts of mutual aid. De Waal observed that low-ranking apes would give the support to those high-ranking apes who were most likely to protect them from tyrannical behaviour on the part of other high-ranking males, and who were most generous in distributing food. Some apes had partnerships, members of which would solicit help from the other partner when attacked, and interestingly, if support was *not* given when legitimately requested then the requesting ape would react with anger towards the unhelpful partner. Capacity

[57] See Morris Goodman, Richard E. Tashian, and Jeanne H. Tashian (eds.), *Molecular Anthropology* (New York: Plenum, 1976) for discussions of the classification of these species.

[58] Howard T. Dene, Morris Goodman, and William Prychodko, 'Immunodiffusion Evidence on Primate Phylogeny', in Goodman *et al.* (eds.), *Molecular Anthropology*, pp. 171–95.

[59] 'Immunodiffusion Evidence on Primate Phylogeny', p. 191.

to win and hold the support of low-rankers seems to be an important factor in deciding which male chimpanzee will lead the group.[60] In another incident de Waal noted:

Tarzan is kidnapped by his 'aunt' Puist. Tarzan is about one year old . . . When Puist is high up in the tree Tarzan panics and starts to scream, so that his mother, Tepel, comes rushing up. Tepel . . . becomes extremely aggressive. When Puist has climbed down again and Tepel has Tarzan safely back, Tepel turns on the much larger and more dominant female and begins to fight with her. Yeroen [one of the dominant males of the group] rushes up to them, throws his arms around Puist's middle and flings her several metres away.

This intervention was remarkable, because on other occasions Yeroen had always intervened in Puist's favour. This time, however, he agreed, so to speak, with the protest of the mother and waived his usual preference.[61]

All these observations very much suggest that these apes could be said to have passed a significant boundary between the states of moral patient and moral agent.

Chimpanzees exhibit complex abstracting and reasoning abilities, and very clearly it is not reasonably possible to doubt that they are genuinely 'subjects of a life', with some awareness of past, present, and future. When given the opportunity to learn various kinds of non-spoken, symbolic codes they have demonstrated an impressive capacity to communicate in language-like ways.[62]

Viki, a chimpanzee raised by Catherine Nissen and Keith Hayes, showed clearly that she thought of herself as a person, not an animal, by placing her own photograph together with pictures of humans when posed the problem of sorting human and animal photos into different piles.[63]

Maurice Temerlin, whose family brought up one of the chimpanzees who were trained to use deaf-and-dumb signs, repeatedly insists that he and his wife both regarded her as their adopted daughter, not as a pet:

[60] Frans de Waal, *Chimpanzee Politics* (London: Unwin Paperbacks, 1982), 207.

[61] *Chimpanzee Politics*, pp. 171–2.

[62] See Ch. 4 for a much fuller discussion of the various ape language projects.

[63] Keith J. Hayes and Catherine H. Nissen, 'Higher Mental Functions of a Home-Raised Chimpanzee', in A. M. Schrier and F. Stollnitz (eds.), *Behavior of Nonhuman Primates*, iv (New York: Academic Press, 1971), 89–90.

[Lucy had created chaos in the Temerlin's house just before guests were due to arrive.] Lucy looked me directly in the eye, smiled her little girl smile, and touched her nose with her thumb, forefinger extended in the ASL sign which means, 'I'm Lucy.' . . . I could not hit her, my eloquent chimpanzee daughter. I think social learning had given her a more effective gesture for appeasing me than evolution.[64]

Genetic relatedness does not necessarily mean that we will automatically be disposed to treat the apes as they deserve. Historically, our own species was probably responsible for the elimination of the other 'men' normally placed within the genus *Homo*. In our mythology creatures which are like but unlike humans—goblins, giants, trolls—are objects of fear and loathing in a way which truly non-human ones are not. It is unfortunate for chimpanzees that this antagonism appears to be readily evoked by their 'goblin-like' appearance. In an astonishing passage Premack and Premack compare the 20-year-old chimpanzee, Sarah, who provided much of the data for their work on concept formation in chimps, to a golem: the medieval mythical creature which Jewish magicians were supposed to be able to create to serve them.

There is a tradition in Jewish mysticism concerning the creation of a golem . . . He cannot speak, but he understands fairly well what is said if commanded. They use him as a servant to do all sorts of housework. . . . While word and picture once competed in the battle for the human mind, the word now stands ascendant. In this book we will once again encounter minds in which a battle between word and image ranges. But here, the outcome is different. This time, in these other minds— in these hairy golem—the word is not destined to win.[65]

Human and chimpanzee differ mainly in their control genes, which comprise a relatively small proportion of the total DNA.[66] Human specialities are achieved by, for example, a delayed cut-off time for brain growth and changes in body proportions. This raises some interesting questions for the future. It is already possible to alter the genetic material of

---

[64] *Lucy: Growing up Human: A Chimpanzee Daughter in a Psychotherapist's Family* (London: Science & Behavior Books, 1975), 183.

[65] D. Premack and A. J. Premack, *The Mind of an Ape* (New York: Norton, 1983), 12–13.

[66] Goodman *et al.*, *Molecular Anthropology*.

some domestic animals by insertion of new genes.[67] These genes are expressed in their new hosts with reasonable success and can be transmitted to the host's offspring by normal reproduction. For technical reasons it is far simpler to insert a new protein gene into the host's DNA and alterations of control genes are still way in the distant future. (How distant this may be is hard to guess, and it should perhaps be noted that only a few years ago a leading geneticist could state categorically that heritable alterations of gentic material apart from random mutation would never be possible.[68] However, the remote possibility that human beings may eventually alter control genes raises important questions. Suppose we were to alter the DNA of an embryonic chimpanzee so that his brain continued to grow after birth, or even to alter all the control genes which distinguish human and ape. Would the resulting child be human or animal? Since chimpanzee and human have 98 per cent[69] of their DNA in common, he could either be said to be genetically 100 per cent human (allowing for some insignificant differences in redundant information) or 98 per cent chimp, 2 per cent human (by inheritance). Or are present-day African apes 98 per cent human by virtue of common inheritance? Presumably it would be possible to modify those same control genes in human embryos to increase the period of brain growth, and brain-weight. One wonders how such an infant would be entitled to view his human progenitors when he became adult. It is already technically feasible to produce human–ape chimeras by fusion of early embryos to produce creatures with a mixture of ape and human cells. Rat–mouse and sheep–goat chimeras have been created in the laboratory[70] and grown to produce viable adult animals with intermediate characteristics. The process does not appear to have caused gross disruption of the behaviour of the chimeras and it is an

[67] See e.g. Robin H. Lovell-Badge, 'Transgenic Animals', *Nature*, 315 (20 June 1985), 628–9, and Robert E. Hammer *et al.*, 'Production of Transgenic Rabbits, Sheep and Pigs by Microinjection', *Nature*, 315 (20 June 1985), 680–3.

[68] Jacques Monod, *Chance and Necessity* (London: Collins, 1972), 153.

[69] J. Gribbin, *In Search of the Double Helix* (London: Wildewood House, 1985), 340–2.

[70] Carole B. Fehilly, S. M. Willadsen, and Elizabeth M. Tucker, 'Interspecific Chimaerism between Sheep and Goat', *Nature*, 307 (1984), 634–6; Sabine Meinecke-Tillman and B. Meinecke, 'Experimental Chimaeras: Removal of Reproductive Barrier between Sheep and Goat', *Nature*, 307 (1984), 637–8.

open question what the mental capacity of an ape–human chimera would be. It has at least semi-seriously been suggested that one way to improve the supply of chimpanzees for research purposes would be to employ female humans as surrogate mothers:

existing techniques would allow a fertilised chimpanzee egg to be implanted into a very closely related and more plentiful species, say *Homo sapiens*. The most impressive of such techniques would wrap a chimpanzee inner cell mass in the emptied blastocyst of a human zygote; the blastocyst—the outer layer of the developing embryo— forms a perfectly compatible human placenta, while the inner cell mass develops into a chimpanzee.

These proposals may seem far-fetched, but all have been accomplished in other species, and it would take very little effort to extend the results to chimpanzees.[71]

If 'Lucy', the 6 million-year-old female hominid discovered in Africa, was a 'man' then it is difficult to say why we do not say that gorillas and chimpanzees are also kinds of 'men'. A definition of a 'man' seems to imply something more complex than mere membership of the germ line of the species *Homo sapiens*.

Modern research into genetics has already made the rigid boundaries between species appear less substantial than they were, and future events may well increase this trend to a point at which we are in genuine doubt about what constitutes a 'man'. Suppose, for example, our descendants decided that they wished to increase the brainpower of 'man' to the greatest possible extent. Some biologists have predicted that the maximum possible brain mass for any animal is 4 per cent of its total body-weight.[72] Thus the brain mass of *Homo sapiens* might be increased to this barrier and no further. Increasing the total body-weight of humans could allow some further increase, but not much because of the sheer physical problems of being a land-based, bipedal animal. Even modern humans experience some physical difficulties as a result of their enlarged heads,

---

[71] J. Cherfas, 'Chimps in the Laboratory: An Endangered Species', *New Scientist*, 27 Mar. 1986, pp. 37–41.

[72] e.g. George A. Sacher, 'Maturation and Longevity in Relation to Cranial Capacity', in R. H. Tuttle (ed.), *Primate Functional Morphology and Evolution* (The Hague and Paris: Mouton, 1975), 417–41.

notably in giving birth, and in walking with a pelvic girdle
which has had to be expanded to allow the passage of a large-
headed human infant. However, a marine mammal would be
subject to no such constraints. Genetic engineering of, say,
whales could produce animals with far more massive brains
and far greater intelligence because the support of their native
element would counteract the damaging effects of gravity.
Hence, our putative 'people engineers' would be able to
succeed to a much greater degree by modifying cetacean germ
lines rather than human ones. The resulting creatures would
be the heirs of human effort, although not of human genes and
I believe that we should be forced to say that, like 'Lucy', they
would be 'men'.[73] There is an obvious parallel with the
problems of classifying extraterrestrial organisms discussed
earlier. A system based upon functional characteristics will be
more useful for some purposes than one based on evolutionary
relationships. It is difficult to define at what point we will decide
that an individual cannot be classified as just an animal, other
than by simply restricting our definition to members of the
species *Homo sapiens*. As suggested earlier, this is unsatisfactory
because we cannot then explain why an impartial observer
should think that the difference between, for example, humans
and rats is any more significant than the difference between
rats and porcupines. There is nothing particularly significant
about the biological means of transmission and continuity of
human germ-plasm. Our cenozoic ancestors were certainly not
human and most probably lacked a conscious mental life, since
they were no more than simple cells or aggregations of cells. A
human grown from wholly synthetic DNA would be no less
human and no less an object and agent of moral responsibility
than you or I.[74] If the 'synthetic' human had modified

[73] The supportive effect of the marine environment perhaps explains why
modern cetacean brains are so well developed in spite of the apparent lack of
powerful selection pressure for intelligence acting upon them. If it is the case
that intelligence is a 'good thing' *ceteris paribus*, but that the physical
difficulties of brain enlargement are greater on land, then much weaker
selection pressure would be needed to force brain expansion in sea animals.

[74] It is already possible to make synthetic DNA by purely chemical
processes, which will code for the production of some individual proteins of
economic significance (J. Cherfas, *Man Made Life* (Oxford: Blackwell, 1982),
182). At present we know the complete DNA sequence for a few of the most
simple organisms, principally viruses (ibid. 114–25), but there is no reason in

characteristics, so that he was not precisely like normal humans, I think we should still say that he was a man, provided that he possessed some subset of normal human attributes, and I believe that, if we accept this, then we must say that *any* organism which possesses this subset of attributes must also be classified as a kind of 'man', although not as *Homo sapiens*. Modern genetics, by opening the way to genuinely conscious control of evolution, and by breaking down the barriers to gene transfer between species, has brought about a fundamental change in the potential relationship between different forms of life. Just as 'Lucy' was a potential step along the road to modern *Homo sapiens* and a fertilized human egg is a potential adult human, so, in some ways, any animal is a potential progenitor of human-like creatures.

When we are faced with an immediate problem of how a particular animal ought to be treated, however, the central question we want to answer is the extent to which that animal is capable of experiencing painful or enjoyable sensations. As this chapter has indicated, basic knowledge of the groups of organisms which exist on earth can give us some broad guidance on identifying those with developed nervous and sensory systems. In the next chapter I will go on to consider in detail what biologists have to say about the possibility of animal consciousness.

principle why it should be impossible to solve the gene sequences of higher organisms. The technology needed to sequence the human genome is already available and biologists are debating only whether it is worth doing given the amount of resources which would be needed. The idea of a 'synthetic' human is not impossible, though unlikely.

3
THE NEW CARTESIANS

... animals do not see as we do when we are aware that we see but only as we do when our mind is elsewhere. In such a case the images of external objects are depicted on our retinas, and perhaps the impressions they leave in the optic nerves cause our limbs to make various movements, although we are quite unaware of them. In such a case we too move just like automata.

René Descartes, *Philosophical Letters*.

... Descartes was as nearly right as makes no matter. If we walk down an English country lane, we walk by ourselves. Trees, birds, bees, the rabbit darting down its hole, the cow heavy with milk waiting at the farmer's gate are all as without insight into their condition as the dummies on show at Madame Tussaud's.

N. Humphrey, *The Inner Eye*.

The Question of Animal Awareness

Consciousness is something of an embarrassment to biologists. The success of neurophysiology in explaining behaviour in terms of the physical activity of brain cells appears to leave no useful function for conscious feelings—if nerve activity is a sufficient mechanism to make me jump aside when a car swerves towards me, what does the startled feeling I have *do*? The theory of natural selection explains the origin of behaviour in purely mechanical terms, leaving no place for 'will', or 'desire', or 'purpose', except in terms of the purely contingent wants of conscious evolved creatures. Genes which produce patterns of behaviour that enhance the survival of copies of themselves increase in frequency, others decrease. Furthermore, consciousness cannot be measured or observed, which distresses biologists who want to ensure that their science is as rigorously objective as physics or chemistry.

Several biologists[1] have commented that it is difficult to explain why even human beings are not entirely non-conscious, protoplasmic computing machines, and that speech itself, which Descartes considered the essential mark of reason,[2] could be as readily produced by such creatures as by conscious people. After all, the brains of biological 'survival machines' are merely complicated computing systems, built and programmed by the physical forces of evolution over millions of years. There is no intrinsic difference between this process and the more rapid and purposeful way in which we build and program electronic computers. If everything that we do could in principle be achieved by purely mechanical events, such as the firing of nerve-cells, it is hard to see what consciousness can add in terms of evolutionary fitness. Since it is effects on fitness which determine whether or not a particular trait will be transmitted, it would seem that there is no reason why consciousness should ever have evolved. One response to this predicament has been the simple behaviourist solution of ignoring conscious experience or declaring that its importance is minimal.

I do not believe that there is a world of mentation or subjective experience that is being, or must be ignored. One feels various states and processes within one's body, but these are collateral products of one's genetic and personal history. No creative or initiating function is to be assigned to them.[3]

There has been considerable criticism of strict behaviourism as a satisfactory theory of mind, and I shall not here attempt a detailed refutation of the behaviourist stance, but will offer a summary of ways in which the theory has tended to affect attitudes towards animals. Behaviourism has played an

---

[1] e.g. Richard Dawkins, *The Selfish Gene* (Oxford: Oxford University Press, 1976).

[2] It appears that Descartes believed that reason is a necessary condition for conscious experience since he states that animals are merely sophisticated automata with no real feeling or emotion, and that we can tell this is so because they lack language (*Philosophical Letters*, ed. and trans. Anthony Kenny (Oxford: Clarendon Press, 1970)) 54.

[3] B. F. Skinner, *Reflections on Behaviorism and Society* (Englewood Cliffs, NJ London: Prentice Hall, 1978), 124.

important role in shaping the way in which the scientific study of animals has evolved. It has also had much wider implications in terms of effects on the attitudes of people who have acquired some knowledge of its principles during the course of a general scientific education.

Methodological behavourists accept that mental events do exist, while ruling that they cannot be considered as part of a scientific study of psychology because observations of such events cannot be confirmed by other observers. Even this has its drawbacks: as noted by Rom Harré, attempts by ethologists and comparative psychologists to produce a vocabulary for describing animal behaviour which has no theoretical implications have merely achieved a descriptive vocabulary which is loaded with the assumption that animals are automata.[4]

The radical behaviourists accept that reports of feelings can have a place in a scientific study of behaviour, but they claim that 'mind' does not exist as such. For example, B. F. Skinner:

what is felt or introspectively observed is not some nonphysical world of consciousness, mind, or mental life but the observer's own body. This does not mean . . . that introspection is a kind of physiological research, nor does it mean . . . that what are felt or introspectively observed are the causes of behaviour. An organism behaves as it does because of its current structure, but most of this is out of reach of introspection. At the moment we must content ourselves, as the methodological behaviorist insists, with a person's genetic and environmental histories. What are introspectively observed are certain collateral products of those histories.[5]

The radical behaviourists' claim is not that mental sensations do not occur (it would be difficult to believe that everyone, including one's self, is an unfeeling automaton), but that they have no significance. Skinner himself makes the peculiar claim that the belief that people's actions can be explained by the emotions they feel is an error of the same kind as that made by early scientists who explained the behaviour of the physical

---

[4] 'Vocabularies and Theories', in Rom Harré and Vernon Reynolds (eds.), *The Meaning of Primate Signals* (Cambridge: Cambridge University Press, 1984), 90–109.

[5] *About Behaviorism* (London: Jonathan Cape, 1974), 17.

world in terms of the desires of material objects; for example, by saying that things fall because they want to reach the earth.

. . . Aristotle argued that a falling body accelerated because it grew more jubilant as it found itself nearer home, . . . All this was eventually abandoned, and to good effect, but the behavioural sciences still appeal to comparable internal states. No one is surprised to hear it said that a person carrying good news walks more rapidly because he feels jubilant, . . . there is nothing like it in modern physics or most of biology, and that fact may well explain why a science and a technology of behavior have been so long delayed.[6]

Nicholas Humphrey's contrasting suggestion that we tend to explain physics in terms of mental events because it is natural for us to predict the behaviour of other people by inferring their feelings[7] seems much more plausible. In fact, it is highly unlikely that Aristotle really believed that falling stones have experiences. A modern sociobiologist would be most affronted by the suggestion that he imagines that 'selfish' genes really have selfish feelings; there appears no good reason to suppose Aristotle was any less capable of creating technical terms from common language than we are. Skinner claims that we should realize that mentalistic explanations of behaviour are not useful or valid because they have not led to any progress in the scientific study or control of human society.[8] I suggest that this claim is wrong-headed because, in his search for evidence of progress, Skinner is going in the wrong direction. If we look at the humanistic arts, we can perceive clear evidence of a historical development in the scope and richness of ability to simulate the activities of human minds. *A la recherche du temps perdu* is possibly not a better work of literature than the *Chanson de Roland* but there is a sense in which it represents progress beyond it. Proust would have been capable of understanding the *Chanson*, but it is doubtful whether a medieval poet could have understood Proust. To Skinner's complaint that we have not advanced our capacity to control human behaviour, we can answer that such behaviour is not static, and that a technology of behaviour cannot be like a

[6] *Beyond Freedom and Dignity* (London: Jonathan Cape, 1972), 8–10.
[7] *Consciousness Regained* (Oxford: Oxford University Press, 1984), 25–8, 36–7.
[8] *Beyond Freedom and Dignity*, p. 6.

technology of inorganic processes. For example, if some humans develop improved insight, which gives them advantages over other people, it is always possible for the second group to improve their own insight too, and one would expect them to try to do so. If Humphrey is correct in thinking that we have an innate aptitude for learning to become 'natural psychologists' and an appetite for knowledge about the human psyche, which we use to manipulate others while avoiding being manipulated ourselves, then any advance in understanding which might enable some humans to achieve better control of others will also tend to enable those others to resist control. It might be true that a modern educated person could baffle one of Homer's warrior chieftains, but this merely means that he can hold his own against members of his own society. As suggested by Robert Gordon, it is possible that our ability to cope with other people's behaviour by 'putting ourselves in their shoes' is not really like using a theory of human psychology at all, if, by a theory, we mean a system of predictive laws.[9] Gordon suggests that, instead, folk psychology consists of the capacity to imagine being another person, and to predict what that person will do by ordinary practical reasoning. For example, 5-year-old children were shown a play in which one actor hid chocolate, and this was later moved to a cupboard while the actor was out of the room. The children were able to predict that the first actor would look for the sweets in the original hiding place. Gordon suggests that they did this by imaginatively restricting themselves to reasoning from the information available to the actor, ignoring their own true knowledge. Thus, folk psychology can effectively predict human behaviour in any novel situation, and it is not to be expected that it will ever undergo the kind of dramatic revision that we find in the history of theories of the physical world. Folk psychology simply has a tendency to become more complex as human social life becomes more complicated and difficult. It is possible for folk psychologists to make wrong predictions if they are mistaken about the nature of the information available, or about the subject's tastes, habits, reasoning abilities, etc., although Gordon suggests that there is reason to think

[9] Robert M. Gordon, 'Folk Psychology as Simulation', *Mind and Language*, 1 (2) (1986), 158–71.

that we do have at least some ability to imagine what it would be like to have different preferences or intelligence.

Part of the argument between the radical behaviourist and the mentalist seems to arise because attention has been focused upon the wrong aspects of psychology. For example, Skinner:

The distinction between a physical and a mental world . . . presumably arose . . . because in the effort to solve the dimensional problem of mental life; there was not enough room in the body for the copies of the world a person seemed to possess. Later . . . a different kind of discrepancy appeared. Were the qualities of images and ideas to be found in nature at all? . . . Light might be a matter of corpuscles or waves, but it certainly did not seem to be a matter of colors; green was not a wave length of light. This was not a serious problem for early philosophers, who had no reason to question the fact that they lived in a world of colors, sounds, and so on. Nor is it a problem to millions of people today, who also believe that they do so. Nor is it a problem for the behaviorist. . . . people see different things because they have been exposed to different contingencies of reinforcement.[10]

And again on the question of free will:

Operant behavior is called voluntary, but it is not really uncaused; the cause is simply harder to spot. The critical condition for the apparent exercise of free-will is positive reinforcement, as the result of which a person feels free and calls himself free . . .[11]

All this seems to miss the point that what really is surprising is not that the human brain can process large quantities of information in relation to stimuli from the external environment; nor that complex organisms tend to act unpredictably; but that there is an experiencing observer. Skinner seems to suppose that human consciousness simply consists of the behaviour of producing verbal labels for different kinds of nervous input from the sensory systems, while animal consciousness consists of the behaviour of making specific movements in response to input from the systems. He claims that introspection is very limited in its usefulness and importance because it only has access to input from these systems and does not 'know' about all of the non-sensory

---

[10] *About Behaviorism*, p. 79.
[11] Ibid., p. 54.

processing which is taking place.[12] It seems that he does this
because he assumes that sensory systems 'just do' produce
collateral subjective feelings, while other neural processing
does not, and that we would have introspective access to the
other kinds of brain activity if only there had been sufficient
evolutionary time for the development of a sensory system
which monitored the brain itself. However, this seems to yield
more questions than ever. Why should sensory systems be so
special? Why don't they just cause mechanical actions of
the body? We know that there are complicated feedback
mechanisms within the brain: what makes them different
from the conscious sensory systems? Furthermore, we know
that sensory input undergoes processing within the brain, so
that what we experience must have at least some relation to
the state of the brain, rather than just the raw nervous input.

If subjective experience is a surface aspect of brain states,
then observing it is still quite likely to be useful in the same
way that observations of material objects are useful in spite of
the fact that they could be validly described in ways which are
quite alien to our everyday experience. (For example, it would
be ridiculous for a physicist to complain that a behaviourist's
observations are useless because what *really* exists is not white
rats and Skinner boxes, but complexly interacting wave forms.)
If we pursue the analogy with physics, there does not seem to
be any good reason why we should not use inferred mental
entities as explanations in much the same way that we make
use of inferred physical entities such as electron shells or
quarks. And certainly, the unverifiability of emotions is not
proof that they cannot cause behaviour. At one point Skinner
complains:

A writer who says, 'The more I read of the early and mid-Victorians,
the more I see anxiety and worry as the leading clue to understanding
them,' is suggesting an explanation of behavior in terms of feelings
generated by punishing circumstances, where the feelings are inferred
from the behavior they are used to explain. He is not claiming to
have any direct information about feelings, and presumably means
understanding what they said and did, but anxiety and worry are
useful clues only if they can be explained in turn.[13]

[12] *About Behaviorism*, pp. 216–17.
[13] Ibid., p. 63.

But I think it is just as plausible to say that, because we are mentalistic entities, we are interested in inferring what it was *like* to be an early Victorian. If this is so the writer's suggestions seem perfectly valid.

The behaviourist rejection of subjectivity is applied perfectly impartially to human or animal (for example, Skinner insists that no human can be said to 'possess' knowledge, but merely to have been acted upon by the environment in ways which modify behaviour). However, even a very conscientious behaviourist probably does not manage to keep it up all the time when he is actually interacting with other human beings. He probably does care whether his family are really fond of him or are only being pleasant in the hope of inheriting his money. Innate tendencies and social experience ensure that we are likely to continue to respond normally to other people's expressions of emotion, even if our theoretical beliefs tend to lead us in contrary directions.

With animals, who do not always command such instant rapport and responsiveness, it is much easier to act out the consequences of a belief in behaviourism. For a start, many species do not express emotions in ways to which human beings readily react. Most small rodents, for example, tend to crouch in response to frightening or painful stimuli and this can easily be mistakenly interpreted as resting.[14]

Secondly, there already exists a tendency to 'explain away' situations where animals are caused pain, not by denying that animals possess *any* feelings, but by claiming that special circumstances mean that something which appears painful does not actually cause pain in a particular instance. This tendency appears to stem from a self-deluding extrapolation from situations where there genuinely are special reasons why pain does not result from actions which, at first sight, appear alarming. As examples, consider the activities of shoeing a horse, debeaking a hen, and hot-branding a horse. All three are regularly defended by assurances that 'it doesn't really hurt them'. In the first instance, we can be sure that this is true, since we know that, when a mistake is made and sensitive

[14] See a discussion of behaviourism and scientific studies of animal welfare in Marion Stamp Dawkins's book *Animal Suffering: The Science of Animal Welfare* (London: Chapman and Hall, 1980).

tissues are injured the horse will go lame and be useless for work. Horses may at first be frightened by the noise and smell of the shoeing process, but they become accustomed to it and will eventually stand quietly whilst the farrier attends to them. In the second and third cases the claims are almost certainly false, since sensitive tissues with extensive innervation are severely damaged.

Because behaviourism has been such an influential paradigm for biological research, it is necessary to make very careful enquiries to be certain exactly what a scientist means when he makes statements about the perceptions or the mentality of animals. For instance, we might want to know whether rats appear capable of awareness of a sufficiently high order for the suspicion that they are capable of forethought. This is important for a consideration of our possible moral duties towards these animals, since some philosophers, for example, Peter Singer,[15] have argued that the wrongness of killing (as opposed to causing suffering) depends upon factors such as whether a being has any conception of itself as an entity existing over time and whether it can anticipate (and thus can be deprived of) future enjoyments. The ability to make sensible choices in pursuit of goals would indicate that a particular being does not live purely in the sensory impressions of the moment (although, of course, this does not indicate that it has any developed idea of self). If we ask someone who is an expert on the behaviour of rats and he replies that, while his rats are capable of learning to run through mazes to reach a food source, they have no awareness and no 'mental map' of the maze, but are merely conditioned to respond to stimuli from their surroundings, we may be tempted to conclude that we need no longer worry about possible duties towards rats. After all, we would not think we might have duties towards a clockwork rat, and we seem to have good authority that the real rats are similarly no more than mere automata. However, if we then go on to ask our experimental psychologist for his opinion on the method by which humans form mental maps and he replies that there is no such thing, we all simply 'follow the territory', making conditioned responses at the 'point of

[15] *Practical Ethics* (Cambridge: Cambridge University Press, 1979), 93–105.

contact with the world',[16] we may start to have doubts about the rats once more. The psychologist's evidence does not show any proof of a morally relevant difference between rats and humans with respect to their capacity for feeling pain or other sensations.

Even if (as Skinner believes) subjective experience is purely a by-product of certain physical actions,

a person can in a sense feel the purpose with which he plays a smooth scale. But he does not play a smooth scale *because* he feels the purpose of doing so; what he feels is a by-product of his behavior in relation to its consequences.[17]

This does not rob it of significance. It still matters to us whether we are happy or distressed. Indeed, this is the very reason why a discussion of subjective experience is relevant to the question of the way we treat animals. If it were true that only overt actions should be significant for us, then there would be no question about the need to take seriously any overt animal behaviour such as writhing, screaming, or struggling. There is a problem only because it is possible for us to wonder whether such signs are actually indicators of mental experiences, and because the behaviour would have no significance if we could be sure that it was not accompanied by distressing subjective feelings.

Some biologists have attempted to escape from the restriction of behaviourism by redefining consciousness in terms of observations, generally either as evident self-consciousness (i.e. ability to reflect about the self) or as knowledge.

. . . if I define self-awareness as the ability to become the object of your own attention, consciousness as being aware of your own existence, and mind as the ability to monitor your own mental states, then it is obvious that these are not mutually exclusive cognitive categories. . . . This is not to deny the existence of feelings which serve to energize behavior in a large number of species, but lacking the ability to monitor such states I would argue that they are mindless.[18]

---

[16] Skinner, *Reflections on Behaviorism and Society*, p. 105.
[17] Id., *Beyond Freedom and Dignity*, p. 205.
[18] Gordon G. Gallup, 'Towards a Comparative Psychology of Mind', in R. L. Mellgren (ed.), *Animal Cognition and Behavior* (New York: North Holland, 1983), 502–3.

Gallup's motivation seems very largely a desire to provide himself with a defence against behaviourist criticism that the concept of mind has no scientific value because it is not empirically testable. By defining consciousness as the demonstrable capacity to reflect about the self (specifically as the ability to recognize oneself in mirrors), he hopes to refute the claim that mind cannot be the subject of scientific investigation because it is an irrevocably private phenomenon. However, it seems to me that in the attempt he loses sight of what is really interesting about consciousness (that we have feelings and are not robot machines), without really succeeding in his own aims. For example, he says, 'In terms of a computer analogy, a mind can be thought of as a subroutine which monitors and modulates particular features of the system at large, and on that basis makes inferences about the operation and disposition of similar systems' (p. 490). However, if this were all that was involved there would be no need for us to have the *experience* of consciousness, or to suppose that other people do. It is perfectly possible to imagine a machine which has been programmed (1) to change its activity on the basis of particular inputs of data and (2) to 'use' information about the changes to predict how another similar machine would respond to the same data. The concepts of feelings or consciousness are not necessary to explain how such a system could work.

J. L. Gould tries to provide an operational definition of consciousness in terms of knowledge.[19] He says that a necessary condition for consciousness is 'an ability to recognize, in a variety of cases, a logical conflict between a token or sign stimulus and the context which it is supposed to represent'. As an example of failure to meet this criterion Gould describes bees ejecting from their hive a live bee which has been experimentally daubed with oleic acid (which normally indicates that a bee is dead, and hence must be thrown out for hygienic reasons). Gould states that this shows that the bees are not meaningfully conscious of their actions. Clearly the bees do not understand exactly what they are doing; the experiment

[19] R. M. Seyfarth *et al.*, 'Communication as Evidence of Thinking' (discussion session), in D. R. Griffin (ed.), *Animal Mind—Human Mind*, Dahlem Konferenzen 1982 (Berlin: Springer Verlag, 1983), 391–406.

has not, however, told us anything about their capacity for subjective experience. For all I know a bee which has been deceived into throwing out a live and resisting worker feels it is awfully hard work. It is more accurate to say we know that a lot of animals are really not very bright (and most unlikely to be self-conscious) but that evidence about their inability to make deductions (such as understanding that a bee who smells dead, but wriggles, must be alive because movement is a more important characteristic of life than scent is) cannot tell us whether or not they have feelings.

The argument against animal consciousness from evidence that they are unable to understand certain situations depends upon the assumption that we can tell whether a being is conscious by investigating whether it can make 'sensible', non-automatic responses to somewhat abnormal situations. It is proposed that this is so because consciousness evolved as a mechanism for understanding the external world. Where there exists no evidence of understanding there is no awareness. Proponents of this position may argue, for example, 'Nesting geese will retrieve "eggs" which are outside their nests even when these are clearly not at all like their own, and they will continue making rolling movements if an egg is removed during a retrieval session. This indicates that geese are automata with no "picture" of what is actually going on.' Drawing such a conclusion firstly depends upon the assumption that consciousness necessarily has a function. It is possible that it is simply a fact that biological nervous systems, on this planet, do generate subjective feelings when they function. If this turned out to be true, there would be no reason to think that purely automatic behaviour was accompanied by any lesser degree of awareness *that it is happening* than 'voluntary' movements. (Of course lack of understanding would still be evidence for lack of self-consciousness, and we might still think that beings without understanding demand less consideration since they have fewer interests and are not likely to fear death or anticipate frightening events in the future.)

Even humans have some behaviour patterns which they cannot alter at will. It must be possible that consciousness might confer increased flexibility in some areas but not in others. This means that an animal may possess evolved consciousness even if he sometimes behaves in a stereotyped way. For example, it has been demonstrated experimentally that people prefer pictures of infants with exaggerated childish features to ones with normal proportions. This is very like the common tendency of birds to prefer outsize eggs to their own ones. Human experience suggests that, if consciousness was advantageous during *some* waking activities, this would lead to continuous consciousness during the waking period, even though some of this time might be taken up by automatic actions, and only a limited section of a particular organism's behaviour might be open to inspection by its consciousness. The suggestion is made by Gould and Gould

that the degree to which conscious thinking is involved in the everyday lives of most people [could be] greatly overestimated? We know already that much of our learned behaviour becomes hard wired: despite the painfully difficult process of learning the task originally, who has to concentrate consciously as an adult on how to walk or swim, tie a shoe, write words or even drive a car along a familiar route?[20]

(Note the similarities to the quotation from Descartes at the start of the chapter.) This seems to be misleading. We are not unaware that we are tying our shoelaces, even though the details of the process may be 'delegated' to non-conscious levels of our brains. It appears that Gould and Gould are confusing consciousness as deliberation with consciousness as simple experiencing of subjective events. In the case of animals with little capacity for learning it still seems possible that their instinctive, fixed-action patterns could be subject to an organizing consciousness which would enable the animal to have some check on where he actually is, and what he is doing at present. The unconscious processing which gives rise to optical illusions (for example) has nothing to do with the intelligence of the central consciousness and everything to do with the way in which lower-level processing is set to operate. This perhaps explains why the egg-retrieving goose behaves as she does: from the point of view of her consciousness what she sees *is* the illusion of an egg, so rolling it into the nest is not really a stupid action.

[20] J. L. Gould and G. G. Gould, 'The Insect Mind', in Griffin (ed.), *Animal Mind—Human Mind*, pp. 266–98.

Evidence from the performance of computing machines shows that consciousness is not specifically necessary for complex or self-referential behaviour: everything that they do can be fully explained and specified in mechanical terms. It could be that consciousness is an epiphenomenon and there 'just is' some level of complexity at which it appears; or perhaps consciousness makes for greater efficiency of processing, or is in some way an integral part of the physical world. However, 'stupid' behaviour would only be relevant to the question of identifying consciousness in an indirect way if any of these possibilities turned out to be true. It might happen to be the case that some level of stupidity indicated such a low grade of complexity that consciousness would not be involved, but there would be no theoretical reason to pick any particular stupid action as an indicator of lack of consciousness because there would be no reason to connect consciousness with the comprehension of any particular event or situation.

Thus, the 'stupid' behaviour of bees and geese cannot be proof that they lack conscious experiences. If low-level processes 'tell' a worker that what she sees is a dead bee, then whether she has a central consciousness or merely more non-conscious processing mechanisms seems unlikely to change the probability that she will perform the actions which are appropriate for a bee in the presence of a dead bee.

The desperate search for an operational definition of consciousness is probably misguided in any case. Operationalism is not really satisfactory even in the case of thoroughly 'hard' sciences like physics because even apparently simple measurements like reading a value on an ammeter can only make sense in the context of theories.[21]

In an attempt to provide an evolutionary explanation of consciousness, Nicholas Humphrey suggests that it is caused by the need to imagine how other conspecifics will react to particular situations. Conscious beings, he claims, are able to predict the behaviour of their fellows in a way in which non-conscious ones cannot, because they can observe their own mental processes by introspection, and use this information to discover how others will react in similar circumstances. He

[21] M. Bunge, *Method, Model and Matter* (Dordrecht: Reidel, 1973), 69–73.

further claims that it is probable that only those animals who have a high degree of 'social intelligence' (maybe just humans and the great apes) are conscious and that consideration of the evolutionary function of consciousness indicates that some animals which behave as if they had feelings are actually non-conscious:

> there were animals ancestral to man who were not conscious ... They were no doubt percipient, intelligent, complexly motivated creatures, whose internal control mechanisms were in many respects the equals of our own. ... They had clever brains, but blank minds. Their brains would receive and process information from their sense-organs without their minds being conscious of any accompanying sensation. ... they acted hungry, acted fearful, acted wishful and so on, and they were none the worse off for not having the feelings which might have told them why.[22]

> In the case of frogs and snails and cod, however, my argument [that the survival function of consciousness is evidence that other humans are conscious] leads me to the opposite conclusion. ... these non-social animals no more need to do psychology than magnets need to do physics—*ergo* they could have no need for consciousness.[23]

> If ... consciousness provides a basis for psychological understanding, and we can guess that a particular animal would have no great need of that, we have good reason to think that animal is not conscious.[24]

In fact, Humphrey would probably expect that a neurophysiological explanation of brain action could provide a detailed account of the computations involved in self-referential prediction. He is not claiming that consciousness is some kind of vitalistic process. But, if this is so, his theory is not really significantly different from any other physicalist account which says that consciousness occurs when brains process information. He has given no reason why only self-referential processing should have this effect. The theory of evolution has been enormously successful in providing explanations for many biological phenomena, and it is precisely this explanatory power which may make us overlook the points beyond which it cannot help us. We may think we can show that consciousness has survival value, but this does not explain how the matter of

---

[22] *Consciousness Regained*, pp. 48–50.
[23] Ibid., p. 37.
[24] *The Inner Eye* (London: Faber and Faber, 1986), 79.

brains is sometimes able to generate subjectivity. Extra-
sensory perception would also have survival value, but we
should nevertheless be most surprised to find an animal who
had ESP, and we would be very puzzled about how it worked.
Why should self-referential physical processes give rise to
feelings, as Humphrey claims? We often need the capacities to
foresee the future, or to read minds by telepathy, but we don't
have either. If the evolutionary explanation tells us anything it
is only a 'why' answer, not 'how'.

Of course evolutionary theory does often provide good
answers to 'how' as well as 'why' qustions, for example, when
it deals with the construction of material organs. Given
knowledge about the chemical behaviour of the various
molecules which are found in living organisms, we can understand how natural selection could lead to the development of
highly complicated organs like the vertebrate eye from simple
beginnings in which early animals responded to light because
it changed the chemical state of reactive pigments in their
bodies. Such pigments might become localized in discrete
spots, permitting the animals to gain directional information
about light sources. (When the spot points directly towards the
light it will be fully exposed, when it is turned away, the effect
will be reduced.) Later, the spots might be protected from damage
by transparent coverings, and these could then thicken and
gain a tendency to focus light, increasing sensitivity and
directionality. Innervation of the spots could then increase in
density, providing information about the patterning of light
and shade within them. Improvements in the focusing quality
of the protective outer covers could then lead to the production
of a genuine image on the proto-retina. This kind of inference
does provide an answer to the question 'How could the eye
evolve?', in much the same way that one might answer the
question 'How is a house built?' by pointing to architectural
plans, builder's manuals, and so on. However, it seems to me
that this kind of explanation is satisfactory only because we
possess prior knowledge about the characteristics of bricks,
mortar, tiles, and other building materials, in outline if not in
detail. Where the characteristics of the starting materials seem
utterly unrelated to what is to be explained there is something
unsatisfactory about theories which point only to the method

by which these materials might be put together (in evolutionary terms, to the survival value of the evolved organ).

If we think complexity of organization of information is sufficient to produce consciousness, can we avoid the conclusion that other kinds of complex physical processes, such as the biosphere of Earth, the universe, programs in electronic computers, Babbage's analytical engine, or ant colonies, are likely to be conscious? Humphrey's claim that we know they are not conscious because they have no evolutionary need for it[25] seems to beg an important question about the relationship between origin and mechanism. A computer may not need to be conscious, but if its construction incorporates the features which explain how (not why) consciousness is produced in humans, then it must be conscious if physical laws are uniformly applicable.

A further problem with Humphrey's proposal is the immense leap which it requires to move from a purely 'mechanical', non-conscious brain to one with fully fledged causal consciousness. On his hypothesis, this has to happen at a stage when brains are already highly developed (at least to the level of social mammals). For such consciousness to have survival value as a means of predicting the behaviour of others it must be causal, i.e. the brain must act as a result of the findings of conscious thought and not merely generate consciousness as an epiphenomenon of its neurological calculation. It is very difficult to see how the leap to causal consciousness could happen, and why the simpler alternative of non-conscious simulation did not develop instead. And if this is so, then the theory does not even give us a satisfactory 'why' answer. To have superior prospects, the newly conscious individual must not only observe his own thoughts, but then perform the intellectual feat of applying his observations to others as a basis for making predictions. Furthermore, since, by definition, the new individual has *different* mental processes from the 'common herd', his predictions will initially have a built-in error factor, which must diminish their usefulness. If human consciousness is analogous to the vertebrate eye, it seems natural to expect that there will exist simpler, preliminary forms among other animals.

[25] *Consciousness Regained*, pp. 181–7.

In some ways, Humphrey's theory seems to be only a more complicated version of the earlier idea that consciousness must have survival value because it enables animals to detect dangerous or injurious events. It does not address the fundamental question of how it could be possible for a phenomenon like consciousness to alter the material functioning of physical objects like nerves. If everything that a person does can in principle be explained by the information-processing function of the arrangement of neurones in his brain (which is a physical thing, even if there is a sense in which an arrangement is not a material object) it becomes immensely hard to see how consciousness can affect our actions at all. Conversely, it appears equally hard to explain how a physical object, even an immensely complicated one, could generate consciousness, which is an entirely different kind of thing.[26]

Humphrey's theory does provide good reasons why animals with feelings should evolve the capacity to reflect about them, but does not seem to explain the origin of the original feelings. There seems something wrong with the idea that feelings could evolve in order to be looked at—why not simply evolve the capacity to refer to what already exists, namely the pattern of neural activity? I suspect that we should look to the physicists, rather than to biologists, for a genuine explanation of the phenomenon of consciousness.

We know that in our own case brains have (however mysteriously) developed the capacity to produce subjective experience in addition to mere information processing. It is *possible* that only humans have this capacity: perhaps there could be a critical mass of brain tissue which is essential for consciousness. This idea, however, involves an essentially arbitrary assumption. If there is no plausible function of consciousness which is shown by us and by no other species, then the suggestion that the human brain has precisely the level of development at which consciousness and sensation become possible begins to seem rather like all the other anthropocentric assumptions, such as the belief that the Earth is the centre of the universe, which have proved to be failures

---

[26] See ch. 7 of S. R. L. Clark's *From Athens to Jerusalem* (Oxford: Clarendon Press, 1984) for a discussion of the unsatisfactory nature of simple biological theories about the evolution of consciousness.

in the past. The only really plausible candidate for a significant activity which marks a difference in kind between human and animal is propositional speech. However, there are already computer programs, such as the MIT creation SHRDLU, which contain models of their own processes, and can generate simple propositional language. This program can answer questions about movements previously made: for example, 'Why did you move the red block?' might be answered, 'Because you told me to' or 'To get at the green pyramid', according to context.[27] Veterinary textbooks sometimes make a point of saying that animal patients have no symptoms of illness, only signs: '*Symptoms* are subjective, that is, feelings experienced by the patient, and, as such, are not applicable to veterinary medicine.'[28] In other words, a human who *says* she has a pain in her hip-joint has a *symptom* (pain); a dog who limps has a *sign*. Whilst it is, of course, true that animals cannot try to describe what they feel is wrong, it is also true that everyone can actually only indicate their symptoms by means of signs. Dr A. may observe that Mrs B. is limping and also complains of pain, while Mr C., the veterinary surgeon, may observe that the dog, D., limps and growls if his hip is manipulated, but the extent to which this gives Dr A. privileged information about Mrs B.'s real state of mind does have limitations. Perhaps Mrs B. is really mainly in need of someone to talk to, rather than pain-killers; maybe the dog limps to regain attention after the introduction of a new puppy.[29] Maybe both of them are really cunning automata. This distinction between signs and symptoms is also made in toxicology: '*Signs* of intoxication are overt and can be observed (e.g. convulsions, elevated body temperature). *Symptoms* are apparent only to the subject of the intoxication and cannot be observed by others (e.g. headache, blurred vision).'[30] It is likely

[27] T. Winograd, *Understanding Natural Language* (New York: Academic Press, 1972).

[28] F. G. Startup, 'First Aid', in D. R. Lane (ed.), *Jones' Animal Nursing*, 3rd edn. (Oxford: Pergamon Press, 1980), 200.

[29] See e.g. Roger A. Mugford, 'The Social Skills of Dogs as an Indicator of Animal Awareness', in Universities Federation for Animal Welfare, *Self-Awareness in Domesticated Animals* (Potters Bar: UFAW, 1980), 42.

[30] V. K. H. Brown, 'Acute Toxicity Testing', in Michael Balls, Rosemary J. Riddell, and Alastair N. Worden (eds)., *Animals and Alternatives in Toxicity Testing* (London: Academic Press, 1983), 1.

that this is not intended to give the impression that there is doubt about the reality of animals' feelings, but this does not mean it does not sometimes have that effect.

If we reject operational definitions of consciousness and the current attempts to explain it in terms of evolution, can biology give us any real help when we need to decide if we are causing other animals to suffer? I believe that it can. Firstly, if we look at accounts of animal behaviour we can see that what is described is explicable in terms of selective value, but sometimes also gives us a distinct impression of consciousness and emotion, rather than merely the motions of non-conscious survival machines. Possibly excessive concentration upon attempts to explain how consciousness could be generated by evolution has made us look at behaviour from the wrong end: an alternative story about evolution could say that it takes the form we see *because selective pressure acted upon populations of conscious individuals*.

Suppose that there exists a population of animals who are conscious, enjoy social contact (such as mutual grooming), but possess a relatively unstructured social system. Suppose further that their environmental circumstances change, with the result that it would be advantageous for them to co-operate in more structured ways—for example, by changing from individual to communal foraging. Clearly, it is possible to make up an account in which an individual who had a tendency to do more social grooming than normal might promote cohesion amongst his immediate family, thus giving them some advantage over the others. Similar behaviour *might* arise in animal automata by mindless reinforcement of tendencies to stay close to the 'supernormal' groomer, if being groomed was already a contingency of reinforcement. Perhaps it is at least more plausible to imagine the change in terms of mental motivations, such as group members feeling more friendly, more willing to share food, tolerate one another's young, and so on. We may be able to obtain some evidence about the existence of consciousness in various animal species by examining whether their behaviour patterns seem, on the whole, to be ones which conscious subjects would be likely to evolve (relying upon mental motivation), rather than ones which might seem more plausible for automata. If the behaviour of a given species is more explicable in terms of likes, dislikes, fears, pains, and so on than in terms of the activities of mindless robots, this is evidence that natural selection was operating upon conscious material during evolution of that species. An example of such behaviour is mourning in geese,[31] which seems easier to explain in motivational than automatic terms. A goose whose lifelong partner has disappeared shows great distress, searches wildly, and seems to lose his normal sense of self-preservation. Clearly it would be more adaptive for a mechanical animal for which stable pairings had survival value to search for a while *without* putting itself at risk, then give up and find another mate. Looked at in terms of motives and mental behaviour, it seems possible that a conscious animal cannot readily have the benefits of powerful emotional ties without the risk of harmful emotional effects when these ties are shattered. Perhaps the correct question is not 'Can this animal do anything that a non-conscious being could not?' but 'Does this animal do anything that a non-conscious one would not be likely to?'

Secondly, physiological and anatomical evidence can help us to decide whether specific acts are likely to cause suffering. For example, in the case of debeaking of chickens we do have genuine problems in deciding whether we are causing pain. We do not possess beaks ourselves, and, at first sight, they look as though they might well be composed of insensitive tissues like our fingernails, or like the horse's hoof. Biological evidence that chickens' beaks consist of an outer layer of horny material covering soft tissue which extends along the length of the beak and is well supplied with blood and nerve fibres, that some of these fibres are of the type generally concerned with the perception of pain in humans, and that these pain receptors are highly active after debeaking adds weight to the common-sense concern already aroused by what we observe during the process (that the cut beaks bleed, and that the chickens struggle and squawk).[32] The evolutionary kinship and anatomical similarity between ourselves and other vertebrates adds further

---

[31] K. Lorenz, *On Aggression* (London: Methuen, 1967), 178–80.
[32] Agriculture and Food Research Council, *Poultry Research Centre Report 1983* (Roslin: AFRC Poultry Research Centre, 1983), 15–16.

weight in favour of animal consciousness to the evidence from observation of behaviour.

The kind of balance we need between assuming that the other animals must be so totally different from us that introspection can give *no* insight into their lives, and treating them as humans in furry coats is perhaps not so very different from the skills we have to employ when we try to 'put ourselves in the shoes' of humans whose circumstances are very different from our own (for example, when we try to predict what very tiny babies will enjoy or be frightened by; when we wonder what it would be like to have been born the opposite sex). Robert Gordon's observation that effective folk psychology depends on the ability to imagine what it would be like to be another person (with his history, knowledge, and so on), rather than simply imagining being one's self shifted into the other's current circumstances,[33] is applicable to the distinction between useful mentalism and sentimental anthropomorphism. The marine biologist Leon P. Zann noted that he could generally predict the position of fish 'cleaning stations', where cleaner fish meet large fish who require grooming and parasite removal, by looking out for any undersea landmark which was 'conspicuous, large or unusually colorful or beautiful [and] *which attracts his attention*' (my italics).[34] Introspection seems to give some useful insight even into such apparently unpromising subjects as fish.

It seems, however, that we are still obliged to say that the central reason why we believe other people and animals have experience is that *we* are conscious and we have an innate tendency to ascribe consciousness to entities which act in ways which we recognize as signs of sensation. Wittgenstein's comment in the *Philosophical Investigations* seems apposite:

Look at a stone and imagine it having sensations.—One says to oneself: How could one so much as get the idea of ascribing a *sensation* to a *thing*? One might as well ascribe it to a number!—And now look at a wriggling fly and at once these difficulties vanish and

[33] Gordon, 'Folk Psychology as Simulation', pp. 158–71.
[34] L. P. Zann, *Living Together in the Sea* (Hong Kong: TFH Publications, 1980), 367.

pain seems able to get a foothold here, where before everything was, so to speak, too smooth for it.[35]

Consciousness and Self-Consciousness

In addition to the capacity to experience sensations, human beings are able to think about the sensations they experience, and we call this ability self-consciousness.

I shall take the terms 'self-awareness' or 'self-consciousness' to signify the ability to form mental concepts about the self as an entity and/or to reflect about one's feelings. Humphrey argues that this kind of definition is a meaningless tautology since all consciousness can only be consciousness about what is happening to the self: 'consciousness (some would say "self-consciousness", though what other kind of consciousness there is I do not know) provides me with an explanatory model, a way of making sense of my behaviour . . .'.[36] However, I think it is true to say that there can be mental processes which go further than simply experiencing feelings to thinking *about* that feeling. In a trivial sense, of course, all consciousness is consciousness of self, but I want to restrict the term here to a capacity to look at what the self is doing, feeling, and so on: the difference between simply experiencing a headache and thinking, 'What a terrible headache, I should never have had that extra pint.' Such reflection does not need to be highly complicated in order to qualify as consciousness of self. Human activities indicative of self-consciousness include, for example, commenting on one's abilities (someone with mild mental handicap might say, 'I'm not as brainy as other people'); reporting sensations ('It hurts *there*'); and so on. Indications of thinking about the feelings of others can also be quite simple, saying that something was done to avoid hurting someone's feelings, for example.

Evidence for a sophisticated ability to think about the self from 'outside' is available for chimpanzees,[37] who show clear indications that they are able to recognize their mirror images

---

[35] Vol. i, trans. G. E. M. Anscombe (Oxford: Blackwell, 1968), pp. 98e s. 284.

[36] *Consciousness Regained*, p. 48.

[37] Gallup, 'Towards a Comparative Psychology of Mind', p. 473–510.

as themselves. Chimpanzees who were already very familiar with mirrors were anaesthetized and painted with odourless, painless markers in positions which would not be visible to them without the aid of mirrors. On recovering, and being offered a mirror, the chimpanzees at once started to finger the marks, something which they did not do if they had no means of examining their reflections. As yet only humans, common and pygmy chimpanzees, and orang-utans have been shown to have this ability.

Significantly, chimpanzees who had been reared in complete isolation were unable to recognize their mirror images however much experience they had of mirrors. Other social mammals, such as dogs, who cannot recognize their mirror images, also act as though isolation interferes with their ability to act appropriately towards their own bodies. Scottish terrier puppies reared in isolation were unresponsive to painful contact with hot radiator pipes, which a normal dog would avoid.[38] This suggests, although it does not prove, that they also possess conscious, social selves, but lack the intellectual ability to understand the possibility of displaced self-images. To recognize an external image as one's self seems to involve a very advanced capacity to think about the self, and to displace the concept of self beyond the physical body. It is possible that a being might be aware of *selves* (within bodies) well before he could perform the steps needed to recognize the significance of mirror images. Interestingly, Gallup states that monkeys seem to recognize and use the mirror images of other individuals (for example, looking in an appropriate direction for the live monkey when an image suddenly appears in the mirror), although they never learn that the mirror also reflects their own bodies.

One indication of ability to think about selves (which must suggest the possibility of ability to think about one's own self) is evidence that a being forms hypotheses about the behaviour of other individuals and his relationships to those individuals in a way which involves more than a mere observation of their physical properties. A being who appears to have ideas about the knowledge which is available to others, or about their attitudes and personality, should possess the germ of an ability

[38] John Paul Scott, *Animal Behavior*, 2nd. edn. (Chicago and London: University of Chicago Press, 1972), 170.

to think about his own behaviour in the same way. Studies of human cognitive ability seem to indicate that this involves more than just a particular level of intelligence. Researchers at the MRC Cognitive Development Unit showed normal, Down's syndrome, and autistic children a simple puppet-play, in which one doll secretly moved a toy marble which the other had left. The children were then asked where the marble really was, and where its owner thought it was. Both normal and Down's syndrome children were able to answer both questions easily, but autistic children (whose intelligence probably exceeded that of the Down's syndrome pupils) could solve only the first question and appeared not to be able to understand the possibility that other people's mental content could differ from their own, and that beliefs can be mistaken if the true facts are not known.[39] These capacities do not seem to develop even in normal children until quite a late stage—most normal 3- to 4-year-olds also found the idea of false beliefs very hard to grasp.[40] It appears that the ability to imagine what other people feel is not really like a sudden intellectual hypothesis that the behaviour of others is explicable if they have feelings similar to one's own. It seems to be a much more automatic capacity which is innate in all normal humans. We know that we are born with the potential to develop certain skills of perception which enable us to make sense of the world. It is reasonable to suppose that we may also have an inborn propensity to attribute consciousness to particular kinds of living things. Differences in intelligence probably alter the effectiveness of our understanding of differences between the outlook of ourselves and others, rather than the basic capacity itself.

Several primate species other than chimpanzees also seem to be able to solve this kind of problem. Dorothy Cheney and Robert Seyfarth discovered that vervet monkeys seem to know enough about social relationships within their own group to make them look towards a particular mother monkey whose infant's calls are being played on a tape recorder. This sort of

---

[39] Leslie A. M. Baron-Cohen and U. Frith, 'Does the Autistic Child Have a "Theory of Mind"?', *Cognition*, 21 (1985), 37–46.

[40] H. Wimmer and J. Perner, 'Beliefs about Beliefs: Representation and Constraining Function of Wrong Beliefs in Young Children', *Cognition*, 13 (1983), 103–28.

thing might be the result of mere conditioned learning, but it looks suspiciously like an ability to 'put themselves in the mother's shoes', or at least to have an understanding that mothers whose babies are crying will be likely to take action of some sort.[41]

Cheney and Seyfarth observed that vervets appear to react to relationships between group members in quite sophisticated ways. If one vervet has a fight with another, she is subsequently more likely to threaten that monkey's close relations, and her own family are also likely to threaten members of the second family. As an example of this Cheney and Seyfarth describe the behaviour of two pairs of monkeys: two sisters and a mother and juvenile. If one of the sisters was attacked by the mother vervet, the other sister was likely to threaten the juvenile. Only vervets over the age of 3 years seemed to be capable of making this kind of generalization about the significance of relationships between family members. Cheney and Seyfarth suggest that being able to do this is of value to the animals because the males transfer to different bands when they mature. Being able to predict what others will do on the basis of information about their family relationships is much more efficient than having to learn by trial and error which monkeys will help each other in fights.[42]

Wild baboons appear to plan their actions on the basis of knowledge of the information which is available to their fellows. For example, a young female was observed to position herself so that only her head was visible to the troop leader when starting to groom an adolescent male (an activity which has a high probability of being punished by the dominant male if he is aware that it is happening). She thus appeared to understand that avoiding punishment depended on her keeping sight of the dominant, while not letting him see what her hands were doing.[43] Chimpanzees again show sophisticated ability to solve this kind of problem, for example, being able to

[41] Dorothy L. Cheney, 'Category Formation in Vervet Monkeys', in Harré and Reynolds (eds.), *The Meaning of Primate Signals*, p. 70.

[42] Dorothy L. Cheney and Robert M. Seyfarth, 'The Recognition of Social Alliances by Vervet Monkeys', *Animal Behaviour*, 34 (1986), 1722–31.

[43] H. Kummer, 'Social Knowledge in Free-Ranging Primates', in Griffin (ed.), *Animal Mind—Human Mind*.

'complete' the action of a film in which a human was shown struggling to do something by indicating what object (out of a choice of two) would enable the actor's goal to be realized. (For example, after seeing the actor shivering and fiddling with a gas heater, the chimpanzee Sarah selected a picture of a lighted spill in a choice between burning and extinguished spills.)

However, when Sarah was required to choose between pictures showing the location of full and empty food canisters after viewing films in which the actor either did or did not see the food location, she did not seem to select on the basis of the actor's knowledge. Instead, she consistently chose favourable outcomes for people she liked, and unsuccessful ones when the actor was someone she disliked. Thus, her behaviour was not like that of either the autistic or the normal groups of children. It does not necessarily prove that she was not capable of making decisions about the knowledge available to other individuals, since she may well have been indicating what she wanted to happen, rather than what was likely, especially as she had previously been required to show what the actor ought to do without reference to his knowledge.[44] Wild chimpanzees do seem to make use of information about the knowledge available to other individuals. Jane Goodall describes how a subdominant chimpanzee was able to preserve sole access to a banana cache by carefully avoiding looking towards it (which would indicate to other troupe members that something of value was located there) until he was the only chimpanzee present.[45] This seems to indicate that this chimpanzee was thinking about his own behaviour: i.e. that his tendency to look towards the bananas would attract the other apes if he did not control it.

Some species of wild cetacea (members of the whale family) have been observed to attempt to aid distressed animals not of their own species. Man is the only other wild animal known to do this, and it is possible that the common factor is an ability to recognize what distress is (rather than simply reacting to

---

[44] G. Woodruff, 'Methods for Studying Cognition in the Chimpanzee', in Universities Federation for Animal Welfare, *Self-Awareness in Domesticated Animals*, pp. 29–36.

[45] *In the Shadow of Man* (London: Collins, 1971), 95–6.

species-specific signals).[46] If this is the correct interpretation then presumably we must conclude that some cetacean species are moral agents in the way that we are (since presumably a cetacean who recognizes that, say, a human swimmer is in distress is then able to make a choice whether to help or not).

The value of relatively unsophisticated self-consciousness in problem-solving has been documented by R. H. Kluwe, who proposes that self-consciousness is valuable as a means by which an individual's own thought processes can be monitored and improved.[47] Kluwe suggests, on the basis of tests in which humans were asked to solve problems while giving a step-by-step commentary, that what he calls 'cognitive knowledge' (knowledge about one's own thought processes, sometimes called 'metacognition') is responsible for eight types of decision during problem-solving:

1. Identification of the problem
2. Checking (success or failure? progress?)
3. Evaluation (is the chosen plan good?)
4. Prediction (what could I do? what would result?)
5. Regulation of resources
6. Regulation of the subject (which problems to choose)
7. Regulation of the intensity (amount of information processed, duration, and persistence of concentration)
8. Regulation of the speed (should steps be added or subtracted in problem-solving or information processing?)

Kluwe suggests that this kind of mental activity is not essentially different from any other type of thinking about events, and that it can be learned. Some human subjects were better at solving problems than others, and these tended to be those people who were also best at describing and understanding *how* they went about problem-solving, indicating that cognitive knowledge does have value in improving lower-order problem-solving. The suggestion that thinking about mental events might be improved by learning raises the possibility that experience may alter the degree to which

---

[46] K. S. Norris and T. P. Dohl, 'The Structure and Function of Cetacean Schools', in L. Herman, *Cetacean Behaviour* (New York and Chichester: Wiley, 1980), 238.

[47] 'Cognitive Knowledge and Executive Control: Metacognition', in Griffin (ed.), *Animal Mind—Human Mind*, pp. 201–24.

animals of the same species are self-aware. Griffin has suggested that an example of a situation where an animal's degree of self-awareness is artificially increased by training is provided by the experiments of Beninger, Kendal, and Vanderwolf.[48] In these experiments, laboratory rats were trained to make different responses to the same signal, depending upon what they happened to be doing at the moment when the signal was given. Griffin argues that perhaps normally rats do not 'pay attention' to what they are doing (but are simply aware of physical stimuli, as they move around) and that the training possibly gave them a new degree of self-awareness and a new ability to relate their actions to external events. Certainly experience does have effects on brain development and problem-solving behaviour in rats, since individuals who are given more complex, 'enriched' surroundings can be shown to develop larger brains and to be more successful in learning and solving problems than ones who are forced to live in more restricted environments. However, it is possible[49] that rats may have a tendency to perform different actions in particular places, so that they were learning to respond according to where they were rather than what they were doing. If so, the training programme may have given them new thoughts about their surroundings, but would not have increased attention to the self.

As Griffin has pointed out, animals must constantly receive information from their bodies, and to assume that they can form *no* concepts about this seems implausible, since information about the self (as a physical entity) must have at least as much survival value as information about the external environment. If animals are aware at all, then it seems that they must inevitably learn something about themselves, even though this may be constrained by their intellectual capacities.[50]

---

[48] Donald R. Griffin, 'The Problem of Distinguishing Awareness from Responsiveness', in Universities Federation for Animal Welfare, *Self-Awareness in Domesticated Animals*, pp. 4–10. R. J. Beninger, S. R. Kendal, and C. H. Vanderwolf, 'The Ability of Rats to Discriminate Their Own Behaviours', *Canadian Journal of Psychology*, 28 (1974), 79–91.

[49] M. J. Morgan and D. J. Nicholas, 'Discrimination between Reinforced Action Patterns in the Rat', *Learning and Motivation*, 10 (1979), 1–22.

[50] 'The Problem of Distinguishing Awareness from Responsiveness', p. 7.

One kind of behaviour which seems only to be explicable in terms of some degree of consciousness of self is the imitation of voluntary actions of others. Some animal imitation is probably explicable in terms of 'social facilitation' (as when one person yawning sets off everyone else in a room doing the same without any deliberate imitation). However, some instances of imitation are much more impressive. Again, chimpanzees show outstanding abilities—for example, the young chimpanzee adopted by Washoe, one of the original sign-language project chimpanzees, acquired the use of thirty-nine signs by imitating her, and without any human coaching. Home-reared chimpanzees have been observed to imitate virtually all of the actions performed by their caretakers.[51] In her review of veterinary aspects of feline behaviour Bonnie Beaver describes an orphaned kitten who was reared together with dogs and developed the habit of lifting his leg at trees by imitation of the male dog.[52] If this observation is accurate the cat's behaviour seems only explicable on the hypothesis that he was aware of the correspondence between parts of his own body and those of his canine companion. It seems likely that such awareness is of the same automatic type as the human ability to think that other people have feelings, rather than an intellectual feat of invention. However, this kind of recognition seems to require some degree of consciousness of self. In *The Concept of Mind*, Gilbert Ryle also suggests that imitation is important in the development of self-consciousness.[53] Ryle suggests that humans learn self-awareness through the performance of activities such as mimicry which involve 'commenting' on the actions of other people and relating them to their own actions. It is particularly interesting that he seems to be suggesting that human self-consciousness is not something which a child can be clearly said to have or lack at some particular moment, but a continuous development which can exist to a greater or lesser extent, contingent upon the child's mastery of the complex skills involved in self-commenting and controlling. Ryle

[51] J. Goodall, *The Chimpanzees of Gombe* (Cambridge, Mass. and London: The Belknap Press of Harvard University Press, 1986), 23–4.
[52] *Veterinary Aspects of Feline Behavior* (St Louis, Mo. and London: Mosby, 1980), 34–5.
[53] *The Concept of Mind* (Harmondsworth: Penguin Books, 1963), 184–5.

suggests that self-consciousness is composed of a battery of abilities (like mimicry, co-operation, retaliation, criticism, reporting, exchanging, persuading, and so on) which children gradually learn. A similar approach to animal self-consciousness would be fruitful. Instead of demanding an answer to the question 'Is this animal self-conscious or not?' it could be more helpful to recognize that species may possess many, some, or none of the complex cognitive abilities involved in self-awareness. Asking the extent of such abilities would enable us to draw up a (very rough) scale on which to assess (for example) the comparative degree of loss involved in killing different species of animal. As indicated above, I am unconvinced by theories of the origin of consciousness in self-reference.

However, Humphrey's argument for the *social* value of consciousness may well hold good for the elaboration of self-consciousness. Self-consciousness might initially take a fairly simple form. It is possible to be aware that what one sees is not what another sees, without possessing elaborate theories of knowledge. It appears that evolutionary pressure for reliable prediction of the behaviour of others would give an important selective advantage to self-conscious animals within a group who were already conscious. Self-consciousness seems to be something which could be expected to evolve gradually (and which probably develops gradually as we grow up), starting from very simple beginnings, like the ability of rats to use knowledge of what they are at present doing as cues to decide the correct response to an unvarying signal, and going on to much more complex calculations about what one would feel in certain circumstances (and hence what the other chap is likely to feel like if *you* manœuvre him into the same situation). Knowing in advance what one would feel in any of a choice of possible situations is presumably likely to have advantages independently of any help it gives in predicting other people's motives. Interspecies 'thought prediction' similar to the intra-species prediction shown by baboons would clearly also have survival value, so I am not certain that Humphrey's concept of a social environment is sufficiently broad. It seems that self-consciousness could have value in any circumstances where conscious beings were interacting (for example, in predator–prey or symbiotic relations). It does seem to be a general rule that highly intelligent, large-brained species (which are the most likely candidates for self-consciousness) are also social ones (for example, elephants, anthropoid apes, cetacea, canids, felines), and that high intelligence seems to be linked more with social living than (as had previously been suggested) with activities like hunting. There does not seem to be any reason except social factors to explain the high degree of intelligence and brain development shown by animals such as elephants and gorillas, who are physically far too formidable to need quick wits to preserve them from natural predators and who live largely on vegetable matter, which is easily obtained merely by stretching forth hand or trunk. Prey-catching or predator avoidance, of course, would be enhanced by an ability to guess what the prey or predator perceives, so presumably this has some effect in providing selection pressure for awareness about thoughts and perceptions. It was suggested earlier that the human ability to attribute thoughts and feelings to others could be an innate capacity, rather than the result of inductive reasoning about the way people and animals behave. If so, it seems reasonable to suppose that it is this capacity which produces the differences between the species which Humphrey thinks are conscious and those he thinks are not.

If some animals are conscious, but less self-conscious, than we are, this does not mean that their feelings must be less intense than ours. It is difficult for us to imagine what consciousness without awareness of ourselves watching would be like. Some idea can perhaps be gained by remembering how it feels to watch or feel with such involvement that we 'lose ourselves' in the action. We forget to think about ourselves watching. Thus the experience of animals may be more rather than less intense than ours. They have less capacity for distancing themselves.

4

ANIMAL COMMUNICATION?

. . . it is evident that magpies and parrots are able to utter words just like ourselves, and yet they cannot speak as we do, that is, so as to give evidence that they think of what they say. . . . And this does not merely show that the brutes have less reason than men, but that they have none at all.

René Descartes, *Discourse on the Method*.

The parrot now employs English vocalizations in order to request, refuse, identify, categorize, or quantify more than 50 items, including objects which vary somewhat from the training exemplars. He has acquired functional use of 'no' and phrases such as 'Come here,' 'I want X', and 'Wanna go Y' in order to influence (albeit to a limited extent) the behaviors of his trainers and to alter his immediate environment. He is also beginning to use the phrases 'What's this?', 'What color?' and 'What shape?' to initiate interactions. More importantly, this subject has also demonstrated the ability to decode our questions reliably: when queried about the color *or* shape of objects which vary with respect to *both* color and shape (e.g., he may be asked the color of a green wooden triangle or the shape of a blue rawhide square), he is capable of determining which of the two categories (color or shape) is being targeted, deciding which instance of what category is correct (e.g. 'green' rather than 'yellow'), and producing the appropriate set of vocal labels 'green wood,' 'four-corner hide'.

Pepperberg, in *Dolphin Cognition and Behavior*.

This chapter discusses the relevance of animal communication studies to consideration of their moral status. Some critics of the idea that animals are the sort of beings which are capable of possessing rights (or even possessing interests) have based their claims upon the proposal that human consciousness stems from our linguistic ability. An alternative objection has been posed by Frey, who does not deny the possibility of animal consciousness, but claims that language is essential for the possession of interests, without which rights cannot be possible. (In fact, Frey does not believe in the existence of rights anyway.[1] However, he has provided an extensive exposition of the claim that animals do not qualify according to the criteria which have been generally accepted as the basis of rights-holding status by people who do believe that rights language has a useful role in ethics.)

Clearly, using language to describe feelings does provide evidence in favour of conscious awareness on the part of language users which is otherwise lacking, even if it still does not make consciousness a certainty (since it is possible that a very elaborate robot might be able to manipulate language in situationally appropriate ways, without actually experiencing the feelings it purported to describe). However, mere use of the word 'sad' does not necessarily give any more assurance about what exactly someone else is feeling than does observation of a sad facial expression. Both inferences depend upon the correctness, for example, of our assumption that the person we are observing is basically similar to ourselves, that they have acquired language by the same kinds of social processes as we have, and so on. Someone with a damaged brain, for example, might repeat words in a highly abnormal way which would not reveal feelings correctly. We may possibly increase the probability of correctness by asking for more information in the form of detailed descriptions of the feeling, and obtaining metaphors which feel appropriate to our own experience, e.g. 'numbness', 'black feeling', and so on, but we cannot eliminate the problem entirely.

It is important to remember the problems briefly discussed above, when we consider the question of animal awareness, and in particular how sure we can be about what animals are feeling and whether it is possible to achieve any form of communication with them. There are difficulties about the interpretation of information about feelings even where humans are involved. A solipsist might even doubt that he possessed

---

[1] *Interests and Rights: The Case Against Animals* (Oxford: Clarendon Press, 1980), ch. 7.

good evidence that other human beings possessed any feelings, but I think it is reasonable to say that this sort of doubt is rather akin to doubts about whether the external world has any real existence. We are obliged to act as though we think the world does exist and other humans do have feelings if we want to act meaningfully at all. The parallel with doubts about the consciousness of other humans is not just a theoretical one. Nicholas Humphrey seriously doubts whether young children possess any inner life prior to the period when they show evidence of being able to think about the inner experiences of other people.

For the very same reasons that we should doubt the existence of consciousness in animals, we should, I think, doubt the existence of consciousness in human babies for at least the first few months of life. ... So far as the baby ... is concerned, there is no evidence that he has an inner life at all. . . . It takes at least a year before children pass either of the tests that I mentioned previously as being applied to animals [recognition of one's own image in a mirror and ability to make guesses about other people's feelings].[2]

It appears likely that certainty about feelings would be extremely hard to obtain in respect of members of another species, even if they proved to be as good at solving problems ('intelligent') as we are. Hence, when making judgements about the way in which we ought to treat such beings, it is not reasonable to demand *certain* proof that a particular practice of any kind causes them to experience suffering before we will abandon that practice. A better way would be to attempt to find out what such beings would choose if they had control of the situation, as in choice experiments in which battery hens were allowed to pick the flooring which was most comfortable for their feet. Still, the idea of communication with animals or of language-like communication *between* animals does, if true, add some extra weight to the belief that animals are the kinds of beings who can have rights; it may be of help in discovering what the interests of animals are; and it is also a fascinating subject in its own right.

I shall consider this topic under four main headings:

---

[2] *The Inner Eye* (London: Faber and Faber, 1986), 93–4.

1. Information transfer. Information about either internal or external factors is transferred from one individual to another. The signal (information) is involuntary and is not intended to affect the second individual, but is simply produced in response to internal or external stimuli.

2. Expression of emotion. Changes in a subject's external state indicate alteration in an internal emotional state, but are not deliberately intended to convey information to any recipient although this may in fact happen.

3. Simple communication. Information about internal states or external events is transferred from one individual to another. This information is semantic: one signal codes one discrete object or event, instead of simply indicating intensity of emotional arousal, and the sender possesses some degree of awareness of the presence of the receiver, in other words, an 'intent to communicate'.

4. Language. Flexible communication system theoretically capable of transferring detailed information about any novel situation and virtually infinitely modifiable. It may have an innate basis in addition to the need for a minimum level of intelligence before a being is capable of using language.

More than one or all of these systems may be present in an individual or species: humans have language; they can also deliberately convey information by facial expression, un-consciously convey their emotions by facial expression, and the kicking of a foetus signals to his mother that he is alive (perhaps disturbed by something), although the foetus can have no conscious awareness of his mother's existence as a person.

It is important to remember that we have no certain evidence that categories 2, 3, or 4 exist in any animal other than *Homo sapiens*. These categories are not those which a biologist would generally use. I have defined them principally in order to illustrate what effects knowledge about animal communication could have upon our beliefs about non-humans.

Consider, first, animals which use only category 1 systems. Examples might be invertebrates such as certain shellfish which secrete biological chemicals (pheromones) which regulate the attachment of members of the same species and so ensure more even spacing. It seems most unlikely that this kind of behaviour would involve any kind of conscious awareness or planning. If we devised ways to 'take part' in such a process of information transfer (for example, by making synthetic pheromone and mixing it with the water in which the animals were growing), we might well succeed in altering their behaviour, but in a way which seems much more like a *physically* controlling action (such as giving a man an injection of adrenaline) than what we normally think of as communication. We can conclude that evidence for the existence of this sort of system is not evidence of consciousness.

A much more contentious case is the dance system by which worker honey-bees 'tell' other workers the location of rich nectar sources. This system is frequently referred to as a language, and does in fact have some features which are more akin to human language than the signal systems of social mammals. (The dance system is symbolic: it refers to objects which are not immediately present by means of arbitrary conventions using the sun as main reference point, represented as vertically up the comb, so that the dancer indicates direction relative to the sun by dance angle relative to the vertical. Distance is represented by the number of 'waggles' performed during each turn of the dance, and different races of bees may use slightly different conventions.)

Bees are capable of indicating both the flight direction and its duration, and of some flexibility if the hive is interfered with so that they cannot dance on a vertical comb. However, there is no evidence that bees can convey information about anything other than the location of food or of shelter at the time of swarming. Hence, although their information content can vary, this variation is of a highly specialized kind. We cannot be certain whether the bee has any kind of awareness of what she does.

Bee 'language' is perhaps usefully compared with the performance of SHRDLU (see Chapter 3), which also has a very small 'brain' (it is allocated only a small section of computer memory to work in) and is flexible within strictly defined limits (for example, it can learn new words, but not new syntactical expressions, from the 'conversations' in which it is engaged). SHRDLU perhaps shows how language is possible for the bee's tiny brain—it is 'programmed' in a strictly limited way so can be modified in restricted ways only (for example, different directions, distances, etc.), while human language is much freer and requires more intelligence: humans can learn different types of syntax, use of tones, etc.

Assessment of the experimental evidence is difficult because such experiments almost inevitably have a tendency to equate 'awareness' with 'understanding'. It is quite possible that an animal might experience what was happening without having any real understanding (as is also the case in situations where they make unintelligent responses to physical situations; see Chapter 3). For example, bees do not learn not to respond to the dance of another bee who is repeatedly providing incorrect directions (because she has been disorientated in some way by the experimenter). Even after fruitlessly searching the indicated area many times the other bees continue to respond to the 'liar's' dancing. But this represents a failure of learning and memory, rather than proof that bees are unconscious during each search period. Interestingly, when dancers gave 'impossible' directions, such as the centre of a lake, foragers still did not learn to ignore them, but neither did they search the lake for flowers. Instead they searched the shore nearest to the spot indicated. This does not necessarily prove that they had any kind of insight as we would understand it, since dancers do fairly frequently give erroneous distances under natural conditions because they seem to judge distance on the basis of effort, hence if a following wind changes to a head wind the foragers will tend to land short of the true destination. This means that it would make sense for foragers to be 'programmed' to stop flying and start searching when their directions take them into an area where no flowers are to be found.[3] It is still quite impressive that a creature with a brain as tiny as that of a bee should be capable of such sophisticated organization of her behaviour. Evidence for the 'mechanical' (or otherwise) nature of the bee's thought processes must probably come from other sources in addition to information about communication systems. For the present, perhaps all we can say is that possibly

[3] J. L. Gould and C. G. Gould, 'The Insect Mind', in D. R. Griffin (ed.), *Animal Mind—Human Mind*, Dahlem Konferenzen 1982 (Berlin: Springer-Verlag, 1983), 266–98.

the almost incredibly complex development of bee society and behaviour demands respect which has an ethical quality, even if bees are not in fact sentient beings.

Category 2 communication is essentially involuntary, but expressive of an internal emotional state. Human beings show this sort of behaviour: we make facial expressions of emotion even when we are alone and do not intend them as a form of communication (although we can also produce some of these expressions voluntarily without feeling the 'correct' emotion to accompany them). This kind of communication is a significant source of information about the interests of animals, and skilled investigation can help us to detect the first signs that animals are becoming distressed or emotionally disturbed. It seems quite difficult to see how to distinguish expression of emotion from semantic reference in animals, at least in any rigorous way. Perhaps we can say only that there comes some point at which explanation of specific communication signals by proliferation of putative emotional states just becomes too implausible. There is evidence that some primate species produce a large variety of calls which are associated with external events.

The question of the possibility of semantic communication between animals is discussed by Robert Seyfarth:

While we readily assume that the noises made by humans can stand for some object or concept, ethologists working with non-human primates have generally begun their work with the quite different assumption that vocalizations are simply a manifestation of some internal, emotional state. Rather than conveying specific information about objects or events in the external world, animal signals are assumed to specify only that an individual is, say, nervous, hungry or excited.[4]

Seyfarth and his co-worker Dorothy Cheney challenge this view, on the basis of their field studies of East African vervet monkeys (*Cercopithecus aethiops*). These monkeys typically live in groups of 10–30 individuals: adult males, adult females, and their young. Each groups holds a small territory, which is

[4] Robert M. Seyfarth, 'What the Vocalizations of Monkeys Mean to Humans and what they Mean to Monkeys themselves', in Rom Harré and Vernon Reynolds (eds.), *The Meaning of Primate Signals* (Cambridge: Cambridge University Press, 1983), 43–56.

defended against intrusion by others. Female vervets normally remain permanently in their natal group, but males generally switch groups at maturity. Interestingly transfer is not random, to any adjacent group, but instead particular groups have a tendency to 'exchange' young males, and groups which have undergone such exchange show reduced hostility towards one another.[5]

The most common vocal sound made by vervet monkeys is a low, grunting noise, made in a variety of situations. Seyfarth and Cheney had earlier discovered that vervet alarm calls varied according to the type of monkey predator sighted, and this caused them to suspect that what the human ear perceived as a single type of grunt might in fact be a variety of different grunts, rather as the English words 'pear' and 'bear' might sound alike to someone who does not know what to listen for. Accordingly, they recorded monkey grunts made in different social situations—'grunt to a subordinate'; 'grunt to a dominant'; 'grunt to a monkey moving into an open area'; 'grunt to another group'. Sonograms (time vs. sound frequency) and power spectra (frequency vs. sound amplitude) indicated that the four types of grunt were indeed acoustically distinct, although even a trained human observer cannot hear the differences. However, playback experiments showed that vervet listeners can distinguish between grunt types. Grunts to a dominant usually caused other monkeys to look towards the loudspeaker. In contrast, grunts to another group caused them to look in the direction the speaker was pointing. Grunts to a subordinate caused vervets to move away from the speaker. Grunts to a monkey moving into an open area caused looking both towards the speaker and in the direction it pointed.[6] Similar-sounding 'chutter' calls of vervet monkeys divide into four types, with small but consistent acoustic differences: snake alarm, threat to a group member, threat to a member of another group, threat to a human observer. (Harris suggests that the threat calls developed from the snake alarm, initially as a way of frightening others by 'crying wolf', then by a process of semantic distinction which retains the 'alarming'

[5] Dorothy L. Cheney, 'Category Formation in Vervet Monkeys', in Harré and Reynolds (eds.), *The Meaning of Primate Signals*, pp. 66–7.
[6] Cheney, 'Category Formation in Vervet Monkeys', pp. 58–72.

association with snakes, but keeps the true warning distinct.[7]
Vervets also appear to give acoustically distinct alarm calls for
leopards, eagles, baboons, and unfamiliar humans, as well as two
other distinct calls—pant-threat and 'aarr'—when threatening
other vervets. The predator alarms seem to have both innate
and learned components. Infant monkeys give acoustically
correct calls, but do not seem to know what species pose
genuine danger; they give the eagle alarm for a variety of
harmless large birds (and even for a falling leaf), the leopard
alarm for any large terrestrial mammal, and so on. Cheney
suggests that adult monkeys may encourage the correct usage
by their tendency to follow correct infant alarms by their own
alarm calls and to ignore infant mistakes.[8]

It seems to me that this capacity to distinguish between
'true' and 'false' alarms means that vervet monkeys must be
considered to be capable of having beliefs and emotions
according to Frey's analysis.[9] We can say that the English
sentence 'There is an eagle' can be replaced by the vervet call
'x' and that the monkey may believe either that 'x' is true or
that it is false. Frey suggests that no animal can be said to be
able to assert anything unless he is capable of lying,[10] and
Harris's speculation about the evolution of threat calls provides
a hint that vervets are indeed possessed of this ability. It is
Frey's contention that the possession of interests (and the
possibility of rights according to a Nelsonian model) depends
upon the ability to have desires, that having desires depends
upon having beliefs, and that having beliefs depends upon the
ability to distinguish between true and false propositions. It
seems to me that vervet monkeys can very plausibly be put
forward as non-humans who succeed in satisfying Frey's
conditions.

Savage-Rumbaugh *et al.* describe the use of gestures by a
small group of pygmy chimpanzees who had not received any
deliberate human tuition.[11] Gesturing was particularly frequent

[7] R. Harris, 'Must Monkeys Mean?', in Harré and Reynolds (eds.), *The Meaning of Primate Signals*, pp. 135–6.

[8] 'Category Formation in Vervet Monkeys', pp. 65–6.

[9] *Interests and Rights*, pp. 86 ff.

[10] Ibid., p. 97.

[11] E. Sue Savage-Rumbaugh, B. J. Wilkerson, and R. Bakeman, 'Spontaneous Gestural Communication among Conspecifics in the Pygmy Chimpanzee (*Pan*

before copulation, apparently because it is necessary for male and female pygmy chimpanzees to reach mutual agreement about the copulatory position they will adopt. (The chimpanzees do make gestures in other contexts, but Savage-Rumbaugh *et al.* specifically analysed copulatory gestures because it was possible to make quantitative comparisons of the association between particular gestures and positions.) They noted twenty-one different specific gestures which could be put into three categories: positioning motions, in which one chimpanzee gently pushed the other in the desired direction; touch plus iconic hand motions, in which the gesturing chimpanzee touches the limb to be moved and then gestures the required direction of movement; and iconic hand movements, in which a chimpanzee gestures without touching the other. They suggest that this kind of complex communication is evidence that pygmy chimpanzees are self-conscious, because they believe that it requires that a chimpanzee must be able to understand a connection between the gestural motions of his own body and the movements he wishes to see another chimpanzee make.

The sort of flexible specificity shown by vervets and chimpanzees clearly does have some similarities to human language. Monkeys also seem to process the elements of their calls in ways which are also rather like human use of sound communication. Spectrographic analysis of the wave patterns of calls indicates that they tend to grade into one another; which is one reason why they were originally considered to represent only a continuum of gradually increasing emotional arousal (for example, the monkey might be startled by a harmless bird and really scared of a predatory eagle). However, tests in which calls were played back to monkeys who then were rewarded for classifying them (by making different responses to different calls) indicated that the calls were perceived *discretely*, i.e. when the 'high' end of one call was paired with the 'low' end of the next the two were classed as different, not the same. In much the same way humans who speak languages which use both 'p' and 'b' sounds hear them as

paniscus)', in G. Bourne (ed.), *Progress in Ape Research* (London: Academic Press, 1977), 97–116.

entirely different letters, although spectrographically their wave forms are rather similar.[12]

We seem to be more readily impressed by call specificity when it emanates from primates than from other animals, and also to react differently to evidence of learned communication behaviour. This may be seen as evidence of 'intelligence' in primates but rather the opposite in animals who are not closely related to humans. The observation that blackbirds, for example, will learn to alarm at any object which other birds appear to be afraid of[13] has been cited as evidence that the call is purely mindless and automatic, not even demonstrating emotion. This seems entirely groundless. In the experiment cited the test birds were allowed to see milk bottles while being exposed to other blackbirds who were apparently giving alarm calls at the bottles, but were in fact reacting to a stuffed owl, which was not visible to the experimental birds. In the circumstances becoming frightened by the bottles does not seem a particularly stupid thing to do; indeed humans very often become afraid of things on the basis of hearsay. In natural situations it is clearly valuable for birds to be able to learn about dangers without the need for direct personal experience of the nature of the danger. It is difficult to avoid the conclusion that suggestions that behaviour is language-like will be more favourably received for some species than others and that because we tend to be biased in this way we must be particularly careful in assessing data about communication. Meanwhile, if we are prepared to agree that some monkeys and apes show behaviour which appears to fall within our original category 3 (simple communication) or perhaps in rudimentary form category 4 (language), there seems no good reason to deny that some birds also display this sort of behaviour. This is perhaps not too surprising since independent evidence about problem-solving ability suggests that some bird species are nearly as good at flexible, 'intelligent' behaviour as primates.

Category 3, simple communication, may be innate or learned according to my definition but its essential feature is its *directedness*. It makes sense only in the presence of a

---

[12] Charles T. Snowden, Charles H. Brown, and Michael R. Petersen, *Primate Communication* (Cambridge: Cambridge University Press, 1982).

[13] Gould and Gould, 'The Insect Mind', p. 280.

recipient, or as an attempt to contact a potential recipient. I suspect that many domestic-animal behaviours, such as a dog fetching his lead or a cat eyeing a closed door, looking pointedly at a human and back at the door, would also fall within this group. Even so humble an animal as a tame rat can indicate the desire for a titbit by tugging at a human's sleeve— an action which is probably reinforced by habit and by reward, but which did not originate from totally random behaviour. The rat does seem to be capable of grasping a connection between humans and food which then leads on to direct activity in a profitable direction. Tame rats also signal displeasure by inhibited bites which do not cause actual injury, possibly suggesting that they do have some idea of the human as a social companion rather than a mere object. It may be that they simply react to cues which prevent them biting family members, but it is still surprising that they can generalize from a small, furry body to the large, pink, and hairless appendages of a human being.

Marler *et al.* believe that they have demonstrated communication between domestic chickens which conveys information about the quality of food offered and whether the caller is willing to share food; and which is a probable candidate for an intentional signal.[14] Adult male bantams were presented with food items which varied in their attractiveness (mealworms, peanuts, peas, and non-food empty nutshells) either when on their own, in the presence of a female bantam, or in the presence of another male. The bantams called rapidly if the food item was a highly attractive one and they were prepared to share it, but at a much slower rate or not at all if the food was low quality and/or they were not willing to share. (Willingness to share was evaluated by finding whether the birds did in fact share the food when the 'partner' was released to approach them.) If no receiver was present the males were less likely to give food calls than in the presence of a female, which Marler *et al.* suggest is an indication that bantam food calls are

---

[14] P. Marler, A. Dufty, and R. Pickert, 'Vocal Communication in the Domestic Chicken', *Animal Behaviour*, 34 (1986), 188–93: I. 'Does a Sender Communicate Information about the Quality of a Food Referent to a Receiver?'; 194–8: II. 'Is a Sender Sensitive to the Presence and Nature of a Receiver?'

intentionally directed towards signal recipients and not merely reflex ejaculations. In the presence of another male, the bantams never gave food calls, and they were unlikely to give food calls in the presence of a female before consuming the food item themselves instead of sharing. Females appeared to get information about food quality from the calls, and were more likely to approach when males were calling to a highly attractive food than to a disliked item. (Food items were placed in a position where they were not visible to the female before she began her approach.) Such evidence of intentionality calls into question the assumption that domestic poultry are unlikely to have much awareness of their existence over time. Since poultry are one of the domesticated species subject to the most extreme restrictions during life and slaughtered in conditions which would probably not be accepted for mammals, this is of considerable importance.

Category 4, language, is unique in its capacity to define virtually any situation in terms of shared, arbitrary symbols. It would be very difficult to believe that any being with true language could be other than conscious, rational, and self-conscious, and it is partly for this reason that attempts to train apes and dolphins to manipulate symbols in language-like ways have aroused so much interest. (Again the ability of one bird, an African grey parrot, seems to have been seen as aberration.)[15]

So far the evidence has been controversial and somewhat inconclusive. It seems that it is only possible to train intelligent and social animals to attempt to use human language systems by establishing some kind of personal rapport. This inevitably means that the tester cannot be fully objective, and the chance of unintentional cueing of the correct response is greatly increased. (It has been argued, particularly in the case of the apes, that the animal subjects simply look out for tiny 'intention movements' on the part of their testers and copy these, generating the signs the testers *hoped* the apes would produce.)

[15] Irene M. Pepperberg, 'Acquisition of Anomalous Communicatory Systems', in Ronald J. Schusterman, Jeanette A. Thomas, and Forrest G. Wood (eds.), *Dolphin Cognition and Behavior: A Comparative Approach* (Hillsdale, NJ: Lawrence Erlbaum Associates, 1986).

One of the most serious objections raised about the ape language experiments is that such attempts are fundamentally impossible, in the way that it would be impossible that there could exist functionally flightless birds who are 'waiting' to be taught to fly by human beings.[16] If an animal's brain has not evolved any given function then it is not possible that this function can somehow exist in a 'latent' state which can be revealed by human instruction, since this latent state, being of no practical use to the creature, could never have been generated by the process of natural selection. Obviously, similar difficulties occur when people makes claims such as 'The human brain is like a vast computer whose possibilities are as yet unexplored', and yet it is true that present-day people are able to perform intellectual feats which would have been quite impossible for their physically similar stone-age ancestors.

I think that the two similar difficulties can be resolved by reformulating the questions involved. If the theory of evolution by natural selection is correct, then our brains' properties can only be accounted for by the extent to which they increased the genetic success of our ancestors. Hence, it is not possible for us to have tracts of 'unused memory banks' which can somehow be switched into action to cope with the complexities of modern existence. However, the requirement of functional value does not rule out the possibility of flexible use of the brain to solve new problems—of course this is precisely the reason why 'intelligent' brains are so successful. But if brain use is flexible, it is possible that we can learn ways to improve the logical 'programming' of thought so that less effort (brain space, time, memory, concentration, etc.) is needed to cope with a given type of problem. For example, the invention of arabic numbers clearly made arithmetic a much simpler process. Some of the brain's 'logic' must be inbuilt to some degree, otherwise we could never get started with the process of grappling with the world, but the flexibility of intelligence must mean that there is always a good deal of scope for improved efficiency of our learned systems of thought.

In some ways, what the signing apes 'say' seems disappointingly lacking in 'otherness', the reported discourses of the

---

[16] M. Gardner, *Science, Good, Bad and Bogus* (Buffalo, NY: Prometheus Books, 1981), 391–408.

gorilla, Koko, do not appear strikingly dissimilar to those of any American infant. But, if it is possible that brain 'programs' can be radically modified by learning, one explanation might be that the signing apes have, to some extent, actually been transformed into a semi-human mental condition. Comparison between thinking and mechanical computing may sometimes be misleading, but one way to look at the signing experiments would be to say that human and ape brains are modifiable universal biological computers, with some 'wired-in' biases. It is possible to 'run' some distinctively human programs on ape computers and in this event the mind of the ape becomes increasingly humanized. This raises the question of whether it is possible for some apes to have been so much transformed that they fall within the range of minds which we would accept as human, and how this ought to affect our attitudes towards them.

In addition, there is some evidence that brains have a limited capacity to increase their actual mass (probably by neural cells developing more complex processes) if they are stimulated by an increased work-load. Clearly this kind of 'exercise effect' would have evolutionary value and hence it is plausible that individuals of intelligent species might under certain environmental circumstances be able to develop to an extent not seen in their historical ancestors.

If language is a mechanism for allowing intellectual work to be done more efficiently (in addition to the obvious communicatory function), rather than an instinct like bird-song, then it may be possible for other animals to utilize these intellectual and logical components of language. It can be seen that some aspects of language structure are reflected in activities which are possible for some animal brains: the idea of patterned syntax which allows the generation of new meaning by rearrangement of known words can be paralleled by simple insight behaviour in rats. If rats are allowed to learn a maze in sections, they can later combine these sections (in fairly simple ways) in order to reach a reward. For example, a rat who knows how to reach point b from point a, and how to get from point b to point c, where there is food, but has never before been faced with the task of reaching the food at c from a, can usually do so without laborious trial-and-error learning of the route.[17] This is analogous to the use of language for simple requests, as in ape signing. The Ameslan 'Gimme banana' can be broken up into 'give', which gets the ape to the point from which it is possible to 'reach' various desired objects, in this instance bananas. Clearly, it must be far more difficult to do this if you are naturally only equipped to use this kind of logic for locomotory decision-making, hence if humans have a predisposition to use syntactical logic for communicating it will give them a tremendous head start over an ape who must learn that this possibility exists. However, the objection that a priori animal use of human language must be as absurd as the idea of a species of birds with the potential for flight who have to wait to be taught by humans does not seem to be valid.[18]

(One possible piece of evidence in favour of a Chomskyan innate 'deep structure' to human language perhaps comes from the way human adults talk to small babies. It is extraordinarily difficult *not* to use these stylized modifications of language—raised pitch, repetition, rhyming, rhythm, etc. If true, this would represent a beautiful example of an 'open' instinct; obviously what we say to babies is drastically dependent on the language we have learned, but we modify this learned tool in a seemingly innate way. Incidentally, the way humans modify language to 'fit' the mental age of people they are addressing could explain how Koko, the gorilla subject of Francine Patterson, may have learned to produce rhyming sequences. If Patterson tends to generate rhymes when she 'talks' to Koko in Ameslan, or when she spoke to her as an infant, it is possibly not beyond the mental capacity of an ape to pick up the fact that humans find rhyming sequences significant, and perhaps that they are a sign of affection. Koko uses Ameslan, but hears spoken English also, and presumably she connects 'word' and spoken sound to some extent.)

Attempts to teach chimpanzees, orang-utans, and gorillas to use modified forms of human language have involved four basic approaches:

1. Attempts to train apes to vocalize spoken English. Of these the best documented is the attempt by Keith Hayes and

[17] Cited in S. A. Barnett, *A Study in Behaviour* (London: Methuen, 1976), 149 f.
[18] Gardner, *Science, Good, Bad and Bogus*, pp. 391–408.

Catherine Nissen to teach the chimpanzee Viki.[19] This established that chimpanzees can, with great effort, learn to vocalize a few English words: mama, papa, and cup. They do not appear to possess the necessary vocal apparatus to imitate human speech successfully, and it is noticeable in films of Viki's attempts at English that she frequently actually attempts to hold her lips in the correct position for speech, using her fingers. This suggests that she understood what was required, but lacked the fine control necessary to carry it out, and that systems in which the ape was directly required to make signs manually would be a better test for the intellectual ability to use symbolic reference systems.

There has also been an attempt by Keith Laidler to teach spoken words to a young orang-utan, which gave somewhat more successful results, although Laidler suggests that the difference was probably not due to a difference between the two species, but a result of his particular methods.[20]

2. Ameslan. Several teams have attempted to converse with young chimpanzees,[21] gorillas,[22] and orang-utans,[23] using American sign language for the deaf, and with conflicting results and conclusions. The essential problems of this approach seem to be the extensive potential for cueing the subjects by natural movements and facial expressions of the trainers. Some critics of the programme, e.g. Terrace,[24] have argued that, even where this does not happen directly, the animals tend to respond by repeating the signs used in questions, albeit in a way which has *some* meaning, rather than generating truly

[19] K. Hayes and C. Nissen, 'Higher Mental Functions of a Home-Raised Chimpanzee', in A. M. Schrier and F. Stollnitz (eds.), *Behavior of Nonhuman Primates*, iv (New York: Academic Press, 1971), 89–90.

[20] K. Laidler, *The Talking Ape* (Glasgow: Collins, 1980).

[21] B. T. Gardner and R. A. Gardner, 'Two-Way Communication with an Infant Chimpanzee', in Schrier and Stollnitz (eds.), *Behavior of Nonhuman Primates*, iv, pp. 117–84.

[22] F. Patterson and E. Linden, *The Education of Koko* (New York: Holt, Rhinehart and Winston, 1981).

[23] H. L. Miles, 'Apes and Language: The Search for Communicative Competence', in Judith de Luce and Hugh T. Wilder (eds.), *Language in Primates: Perspectives and Implications* (New York: Springer-Verlag, 1983), 43–61.

[24] H. S. Terrace, 'Apes who "Talk": Language or Projection of Language by Their Teachers?', in de Luce and Wilder (eds.), *Language in Primates*, pp. 19–42.

new statements. Properly controlled tests, in which chimpanzees and gorillas were asked to name objects which they could see but the tester could not, show that these apes could certainly associate gesture and object correctly without prompting, but this of course is not language and a variety of animal species can perform similar tasks (although it is possible that we ought to recognize that all species which are capable of such 'labelling' tasks possess one important prerequisite for language-like behaviour).

Testing for the ability to generate structured sentences (with 'grammar') in order to communicate propositions in an Ameslan system almost inevitably requires human participation. Since apes and humans share a common repertoire of natural body expressions, and since all of the ape subjects have been reared by humans from an early age, this means that there is a constant flow of information between human and animal which is quite independent of the symbol system. This problem is known as the 'clever Hans' phenomenon, so-called after a horse called Hans, who could apparently solve arithmetical problems. Hans in fact gave the correct answers by tapping with one forefoot while closely observing the surrounding humans and stopping as soon as he observed minute involuntary movements which they tended to make when he reached the correct number. If the observers did not actually know the true answer, then Hans was unable to perform. Social animals seem to use this kind of observation quite generally amongst themselves, and they may find it much easier and more natural to transfer the same kind of process to their interactions with humans, rather than learn symbols. This does not prove that they are not capable of anything more than automatic programmed responses to stimuli, after all, human intention movements are not very similar to those of a horse. Hans had to *learn* what he needed to observe in order to solve his problem correctly. It does mean that when humans attempt to establish whether animals have the intellectual ability to manipulate artificial symbol systems in order to generate novel statements (as when someone who understands 'cat on mat' and 'pen in jar' is able to combine the two to say 'pen on mat' without ever having seen this proposition before) the animal will always be liable to extract the solution direct from his testers. In the above example, for instance, the solution might be cued if the tester had a tendency to look from the pen to the mat: it would not be too difficult for a chimpanzee who already understood the association between 'pen' and a sign and 'mat' and another sign to get the correct word order, 'pen—mat'. In fact if the ape did possess the capacity to understand true sentence generation by means of learned grammatical structures he might never have any incentive to learn to do so.

To add further complications, humans are also capable of interpreting non-verbal postural and facial communication and there is probably a tendency for the testers to guess what the ape really 'means' even when he has not succeeded in encoding this correctly. Ameslan (and normal human spoken communication) rely to a degree on non-verbal messages to make sense of what is being said (questions particularly often depend entirely upon tone of voice or facial expression to distinguish them from statements of fact). Once ape and human know one another well and use of signs for objects has become fluent it will be very hard for participants in the exchange to be sure whether information is being transferred by a symbolic grammatical pattern, or if in fact the apes develop an extensive vocabulary but use it in a way which does not really differ very much from signalling by domestic pets.

These factors probably explain why the controversy between instructors who are certain that 'their' apes do have language and the more sceptical is at times so bitter. I doubt that it is coincidental that the trainers who most fervently support the idea of ape language are those who have had close contact with their subjects from an early age and know them extremely well. The most notable of these is probably Francine Patterson, whose two gorilla subjects have lived with her as their primary trainer from babyhood into adolescence. These two apes appear (on Patterson's evidence)[25] to be the most developed language users so far. Critics, however, have accused Patterson of loss of scientific objectivity and anthropomorphism.

On the one hand, one would expect that optimum conditions for teaching language to apes would be a stable relationship

[25] Patterson and Linden, *The Education of Koko.*

with a few gifted trainers in parental roles and contact with other young apes to allow social play and use of language *between* apes. The chimpanzees who performed 'worst' as language users do seem to have been the individuals who were treated as intellectual problems rather than as social beings with feelings and were 'worked on' by a shifting population of graduate students without the opportunity to establish stable social relationships. The 'best' language users seem to have had one major trainer plus helpers. On the other hand, humans have feelings too. It is probable that trainers who have invested emotional, as well as intellectual, capital in a young ape will tend to give their 'adoptee' the benefit of the doubt whenever possible, and will also be the most susceptible to non-linguistic cueing, which gives a misleading ability to predict what the ape will do next. Such trainers may well have a unique insight into the gorilla's or chimpanzee's universe without being the most suitable people to give an objective opinion on whether the apes' proven abilities constitute true language. It is possible that Ameslan itself, with its use of natural gestures and expressions as well as arbitrary ones, is not the best method for investigating this question. In any case there does seem to be a need for a more objective means of assessing the purely intellectual aspects of ape language use. It would be interesting to find out what would happen if an independent observer were able to analyse and test communications between trainer and subject. So far this seems to have been done only in a minor way, for example, to test for the clever Hans effect in naming of objects, and I do not know of any instance where an independent and sceptical scientist has been given a free hand to devise tests of sentence production and comprehension from a viewpoint external to the relationship between ape and trainer.

Language-like communication is essentially an interactive process which takes place between human and ape. Roger Fouts, Washoe's principal trainer, describes signing by Washoe's adopted infant:

he imitated Washoe's signing *hat–George*, . . . Later . . . he was starting to use signs in their correct context. For example, he now will point to an apple, signing *that*, and sign *gimme* over and over again.

He has also been observed to use the *hat–George sign* to refer to all persons—chimpanzees or humans.[26]

Clearly Fouts's concept of the human–chimpanzee relationship is very different from that of some other researchers, who consistently refer to 'the animal' and 'it'.

3. The Premacks have trained their subject, Sarah, a female chimpanzee, to use small plastic symbols to represent 'words' and to respond to strings of symbols by placing tokens on a response board or by carrying out appropriate actions; for example, she responds to the string, 'Sarah insert apple pail banana dish' by placing an apple in the pail and a banana in the dish. This indicates that Sarah, at least, is capable of understanding that word order has significance and can perform without human cues (since a statement can be revealed to her when she is alone, being observed by someone she cannot see). Sarah can respond to 'if' sentences in a sensible way, and she also shows displacement: given the *symbol* for apple she can describe it using symbols for 'round' and 'red' even though the 'apple' token is itself neither circular nor red. Obviously this method has none of the potential for spontaneity of Ameslan and Sarah is restricted in her set of possible answers by the tokens provided. She has not been observed to construct novel compound words to describe objects which have not been named to her as some of the Ameslan chimpanzees have done (for example, 'cry-fruit' for onion).[27] However, she was seen to play at 'questions and answers' with her plastic 'words': putting them into strings which made questions like those she had been asked, and then answering the questions.[28] She was able to learn the meaning of new 'words' by the context of their use; for example, to instruct her in the meaning of the symbol for 'brown' she was given the construction 'brown colour of chocolate'. Since Sarah at that time knew the meaning of 'colour of' and 'chocolate' she was subsequently able to make correct use of the symbol 'brown'.

---

[26] R. S. Fouts, 'Chimpanzee Language and Elephant Tails: A Theoretical Synthesis', in de Luce and Wilder (eds.), *Language in Primates*, p. 72.

[27] M. K. Temerlin, *Lucy: Growing Up Human* (London: Souvenir Press, 1976), 120.

[28] D. Premack and A. J. Premack, *The Mind of an Ape* (New York: Norton, 1983), 29.

None of the other chimpanzees tested on this system has so far proved as successful as Sarah, which suggests that it is just possible that some of the failures to demonstrate correct use of Ameslan were partially due to individual variation between apes. It does not appear to be reasonable for workers to dismiss the reports of successful language training on the basis of one or two failed attempts of their own (although Terrace has also criticized the data produced by workers who claim success on the basis of comparison of their written data with video recordings of the same 'conversations' with the apes, and he argues that the behavioural evidence for language use in these sequences is not conclusive).

4. The Lana Project. Several chimpanzees, the first a female named Lana, were trained to press keys to produce strings of symbols. The project was designed to investigate chimpanzees' ability to produce combinations of symbols and to make it possible to eliminate observer cueing. The keys pressed by the chimpanzee could either be viewed by a human trainer or they could cause a computer to do a variety of things such as turning on a slide projector or releasing a food reward. The chimpanzees trained successfully used the keyboard to make requests and to obey commands, but they seemed not to learn to understand the symbols as *names* of objects in the way that Sarah appears to. For example, Lana was able to use the symbols to request various objects, but found great difficulty in learning that the same symbols could be used to answer the question 'what's that' when the objects to which they referred were displayed to her.[29]

Language training with dolphin subjects is at a much less advanced stage. J. C. Lilly initially attempted without success to train dolphins to imitate human vocal speech.[30] Again, it seems possible that one of the barriers to communication was simply that the environment in which dolphin and human interacted was so barren of significance that the two simply did not have anything realistic to try to communicate about. Initially Lilly's dolphins were confined to small swimming pools. Later he attempted to provide enriched surroundings by

---

[29] E. Sue Savage-Rumbaugh, *Ape Language: From Conditioned Response to Symbol* (Oxford: Oxford University Press, 1986), 8–9.

[30] *Man and Dolphin* (London: Gollancz, 1961).

keeping a dolphin and his trainer together in an adapted house with flooded rooms. However, this still did not offer anything like the complexity of life in the ocean and a distressing number of his dolphin subjects died during the course of the experiments because of illnesses induced by the constraints on their ability to move around. More recent research has been similar to the experiments with apes, but has concentrated on training dolphins to *respond* to sequences of tokens, hand signals, or aural commands.[31] These have indicated that dolphins can generalize from particular use of words to understand new sentences: for example, by obeying the command 'pick up' the first time it is combined with previously learned object names.

Some whales produce complex 'songs', which vary from year to year and whose purpose is not known but is surmised to be connected with the attraction of females[32] (although it is not yet certain that only males sing). Experiments in which captive dolphins were required to communicate with one another in order to solve problems indicate some natural vocal communicative ability but one which falls far short of human language use (assuming that the experiment didn't happen to be the equivalent of caging an English-speaker next to someone who spoke only Chinese and deciding that their inability to communicate proved that humans cannot use language). There have been some attempts to record the natural communication behaviour of wild cetaceans. These seem to indicate that their calls have the same kind of 'graded' structure characteristic of primates and therefore may represent few or very many different 'labels'. The difficulties of linking specific calls to events are of course very great for free-living cetaceans, whose group members may communicate over distances of several miles, and it is probably safest to say that we know virtually nothing of what is, or is not, possible for them.

We can look further at the evidence that apes do possess the level of intelligence necessary for humans to use language. Terrace argues that chimpanzee 'sign language' is very largely

---

[31] L. M. Herman, 'Cognitive Characteristics of Dolphins', in L. M. Herman (ed.), *Cetacean Behaviour* (NY and Chichester: Wiley, 1980), 409–21.

[32] R. Payne, *Communication and Behavior of Whales* (Washington, DC: AAAS, 1983).

imitative and prompted by the apes' human tutors.[33] He seems to be correct in his assertion that most recorded chimpanzee 'sentences' of more than two words are created by addition of redundant signs (such as the addition of the ape's own name, 'hurry', or 'please') for emphasis. However, the example of language prompting which he gives from his own research with the chimpanzee Nim[34] is interesting. In a sequence of video stills, Nim signs 'Me hug cat', prompted by his teacher, Susan Quinby, who signs 'you' while Nim signs 'me' and 'who' while he signs 'cat'. While he signs 'hug' she holds her hand in the 'n' configuration, a prompt for 'Nim'. Now this is not 'monkey see, monkey do' imitation, or clever Hans behaviour, but something very much more like human language (although 'you' and 'me' are very clearly related in Ameslan; 'you' points towards the other person and 'me' towards oneself and it is plausible that there was an element of imitation here).

Possibly the most accurate interpretation of the primate language investigations is that primate signals have an element of openness through a capability for modification by addition, deletion, and restriction of new 'words', while true human language has two sources of openness: word modification and sentence generation.

The chimpanzee Sarah was shown to be capable of reasoning tasks in which she was 'asked' to complete an analogy of geometric figures (A is to A' as B is to either B' or C).[35] For example, when A was a figure marked with a dot and A' an unmarked similar shape Sarah chose an unmarked crescent shape in preference to a marked crescent shape when shown a marked crescent of a different colour as B, in spite of the fact that the 'C' shape (marked crescent) was the same colour as A and A'. She was also capable of correctly classifying pairs of $A : A'$ and $B : B'$ as 'same' or 'different'. She could select the correct completing object in problems of the form: Paper (A) is to scissors (A') same as apple (B) is to either knife or plate. However, she was not able to complete analogies of the form

---

[33] H. S. Terrace, ' "Language" in Apes', in Harré and Reynolds (eds.), *The Meaning of Primate Signals*, pp. 179–203.

[34] Who was later sold (against Terrace's wishes) for toxicological research.

[35] D. J. Gillan, 'Ascent of Apes', in Griffin (ed.), *Animal Mind—Human Mind*, pp. 177–200.

$A : A'$ same as ? by selecting B and B' from a set of six alternatives and putting them in the correct order.

A second chimpanzee, Sadie, was capable of inferring the relationship of non-adjacent pairs in a series of stimuli $A < B < C < D$ where $<$ signified 'less food than'. After training on adjacent pairs Sadie chose the correct (more food) member of the pair $D : B$ on all twelve non-adjacent trials. Chimpanzees were also capable of deducing that food had been removed from a particular location from the sight of another chimp returning from that general direction carrying food, or of another chimp heading in that direction and later possessing food, or of another chimp holding the particular kind of food which had been hidden on the site. (They indicated this by going to the alternative baited site.) This kind of reasoning ability varied considerably between individuals, some appearing much 'brighter' than others.

It is significant that all the apes tested, with the possible exception of the two gorillas, seem to use language much less effectively than would be expected from their ability scores on tests of more general 'intelligence' (although these tests also are probably misleading because they only measure aspects of intelligence which humans think are important). Sarah, for example, is capable of understanding that liquid *volume* is conserved when container *shape* is changed (pouring from a squat container to a long, thin one), something which human children cannot easily do until they are 7 or 8 years old. Her use of language, however, is only as good as a human 3-year-old. Possibly humans have an innate disposition to learn structured symbol systems, and it is even possible that the basic structure of the symbol system is partially innate. If this is so then humans may find language use rewarding in itself while the ape finds it valuable only as a tool to achieve other ends. If human language has innate components then human beings will require *less* intelligence to master it than a comparably competent ape. This is supported by the fact that severely retarded humans often can produce language to some extent and may do so in ways which do not seem to depend on external rewards. Similarly, young children seem to enjoy repeating 'nice-sounding' words and playing with words. It is possible that there may be a genuine human urge to imitate sounds and rearrange them creatively for which there is no analogue in the ape mind. This perhaps explains why the parrot performed surprisingly well in similar experiments: these birds also seem to find manipulating sounds interesting, so the bird who was tested co-operated and concentrated on the tests. In contrast most descriptions of the ape experiments suggest that the subjects often found the constant repetition of naming tests boring and frustrating, and that as they matured they became more likely to try to get what they wanted by force instead of going along with the human language game. It is also evident that the apes very seldom were in situations where they actually *needed* to convey or receive information. One language subject, the chimpanzee Lucy, was subsequently returned to a semi-wild chimpanzee group in Africa and it would be very interesting to know whether she attempted to make more use of symbolic communication in a situation where she actually did need to learn from the people who were rehabilitating her. Intuitively one feels that this sort of communication would be more significant than the captive apes' continual bargaining for candy and cigarettes. In this context it is interesting that one researcher who was studying captive orang-utans was able (by means of natural gestures) to convey to a female ape that he wanted her to retrieve from her mate a piece of dangerous equipment which the latter had stolen.[36] This suggests that these animals *are* able to understand human purposes to a surprising degree when it is obvious to them that the human involved is really alarmed. Possibly the strongest piece of evidence that chimpanzees can learn to see sign language as a means of communication, rather than merely as 'trick' which will enable them to obtain desired items, is Fouts's study of Washoe's training of her adopted infant Loulis.[37] Fouts reports that Washoe actively demonstrated signs to Loulis. For example, she was seen to mould his hand into the 'food' sign while the two apes were watching food being prepared for them. On another occasion, Wahoe placed a chair in front of her infant and repeatedly made the sign 'chair–sit'. Fouts believes that she taught Loulis to understand the sign

---

[36] H. Markowitz, *Behavioural Enrichment in the Zoo* (New York: Van Nostrand Reinhold, 1982), 180.
[37] Washoe's own infant died when 2 months old.

'come' by stages. First Washoe would sign 'come' and then go and pick Loulis up; later she would make the sign, approach, but not actually lift him, and finally she would simply sign 'come' while looking towards the infant. Just over a year after his adoption by Washoe, Loulis had acquired a vocabulary of seventeen different signs, at least ten of which must have been learned from Washoe or the other signing chimpanzees in his group since the human caretakers limited themselves to seven signs in his presence ('who', 'what', 'want', 'which', 'where', 'sign', and 'name').

Because of the strong possibility that human linguistic communication has innate structural components and that humans have a natural desire to manipulate language, experiments in which attempts are made to train other animals to use language need to be regarded with caution when considered as a means of estimating intellect. It is *possible* (though not very likely) that only beings with language can be self-conscious, or perhaps even conscious. Communication experiments are perhaps more interesting and significant as part of a general attempt to explore how animals see the world, than as a kind of test of a particular species' position on some scale of value. They indicate that many species do more complex things than we had credited before, and that we are not the only species able to make use of symbolic representation of the external world. If human language is unique this does not necessarily mean that communication with other animals is impossible, and indeed in some senses it makes humans seem less special if our abilities are not simply the result of superior general intelligence.

If individuals of other species can use human language to some extent through grasping its logical principles, must we then agree that they 'have language' and that the evidence for their moral status is as good as that in favour of a comparably able human? I think the answer is a cautious yes, with the reservation that it could be the case that one part of the inborn nature of human language use involves the connection of logical concepts to mental impressions. It is just possible that we are 'wired' so as to get a mental 'thought' of 'I' when we use the word, while a non-innate user merely manipulates the word in an automatic, 'conditioned' way without having any thoughts to go with it. But if this is so, there seems to be no reason why we should take natural language use by a different species as evidence of awareness of self-consciousness. Possibly in their case the innate programming is automatic and unconscious, a mere sophisticated machine. If we refuse to accept the evidence from behaviour and from similarities of those parts of the nervous system which mediate emotions in humans and the corresponding parts of animal brains, then language does not seem to provide the answer to our demand for absolute proof; as indicated earlier I suspect that there is no kind of evidence which ever could provide this kind of absolute certainty.

In considering human investigation of animal communication, we need to make a distinction between knowledge which enables us to use such communication (for example, to predict what an animal will do next) and knowledge which enables us to analyse the communication system. People who work with animals often 'know', in the sense of being able to act on, the significance of minute changes which would not be apparent to a casual observer and which they find difficult to describe in detail. I suspect that in these cases there is a kind of 'reverse clever Hans effect' in operation, with the human 'plugging in' the animal's behaviour to the normal systems by which we cope with signs and signals from other people. In this case, appropriate feelings and beliefs would be evoked by the animal signals *as if* they were generated by another human. The human participants may not be able to analyse what happens any more than most people could give a detailed linguistic account of a normal verbal conversation.

Scientific analysis of animal communication enables the various elements to be categorized and studied, bringing out aspects which would not normally be apparent to people who use animal communication systems, in the same way that linguists can expose aspects of human language we did not previously suspect. (Of course some individuals can use both methods where appropriate for their needs.) However, during the process of analysis, presumably the normal communication systems 'knowledge' is not utilized, so that animal systems do *not* automatically call up emotional responses, and this uncoupling is reflected in the terminology used (for example, 'vocalizing' rather than 'yelping' or 'whining'). If we only use the second system it may appear more natural to suppose that animals are not conscious, while when we use the first we cannot believe that they are not inhabiting a shared world of subjective experience.

This may explain the antagonism which often exists between people who know a great deal about animals, but have used different methods. Such animosity seems a pity, since the two approaches should complement one another. (When we set out to learn a foreign language nothing will substitute for interaction with native speakers, but a little formal grammar is useful too.) The natural systems of humans have developed over millions of years as a superb means of obtaining information about other living beings. If we refuse to use them as being too subjective, we are refusing a significant source of knowledge. This sort of entering into the social life of animals has often been described as 'empathy'. However, I prefer not to use the term because of its connotations of guesswork or hunches. Using information from the natural communication-processing systems of our brains is in principle no more irrational than using our visual-processing systems to convert light patterns from a fuzzy, meaningless jumble into significant events. Interestingly, signals which are not at all like those produced by humans or by common domestic animals seem to 'plug in' to this system quite satisfactorily. Angry mice perform a display known as 'tail-rattling', in which the tail literally produces a sound like a rattle. Having learned the significance of this action, I believe I now see it directly as 'rage', rather than having a feeling like 'I observe that the mouse is acting in a characteristic way—I therefore conclude that he is likely to bite if I continue trying to pick him up—I therefore desist'. As Stephen Clark has pointed out[38] we see one another's emotions, not probabilities of future actions or behaviour patterns, and this is no more improbable than our ability to see tables and chairs when reflected light sets up electrical excitation of our retinas.

---

[38] S. R. L. Clark, 'Awareness and Self-Awareness', in Universities Federation for Animal Welfare, *Self-Awareness in Domesticated Animals* (Potters Bar: UFAW, 1981) 12.

Analysis of animal communication can show us things we did not notice using our natural systems. It provides a means of looking at the components we could not notice when we were looking at the whole. It sorts out what is false in our beliefs about what we understand and what the animals understand. It may enable us to see communication which is too different from our own to fit the natural systems, or which is outside our normal sensory range.

Premack and Premack suggest that apes who have been taught to use symbolic 'languages' have not learned human language [which seems to be an innate instinctive 'drive' for a species-specific type of learned behaviour] but that they have been enabled to communicate with us in symbolic code.[39] They point out that brain-damaged children who lack the speech areas of their cerebral cortex can often be taught to use similar codes: again, they have not acquired human language, but they have gained a very valuable intellectual tool which can immensely increase their chances of an independent life.

Jane Goodall suggests that the pressures imposed by captivity may sometimes cause primates to develop cognitive abilities which are more sophisticated than those of wild animals.[40] In the wild much energy must be expended on food gathering and processing, but in laboratory conditions testing procedures (such as training in sign language) can focus all the animal's efforts on one cognitive skill. Where primates are confined in social groups, again the elimination of the need to forage allows them to devote themselves to improving their position in the social hierarchy. Because captive conditions do not permit the option of leaving the group when, for example, the levels of aggressive tension become too high, there is a need to develop new strategies for reducing conflict, such as female mediation between rival males.

The results of ape language studies may have something useful to tell us about the ape's point of view even if true language is not involved. For example, use of signs to make humans do things is an effective indication of what is important to the apes; the fact that one of the first signs Washoe learned

---

[39] Premack and Premack, *The Mind of an Ape*, pp. 124–40.
[40] *The Chimpanzees of Gombe* (Cambridge, Mass.: The Belknap Press of Harvard University Press 1986) 583–6.

was that used to initiate social play tells us something important about her.

Perhaps it is possible to find another useful parallel in the language-like codes which have been proposed for attempts to communicate with intelligent extraterrestrials.[41] Such codes are necessarily highly mathematical in content and it seems to me that it is doubtful whether most uninitiated human observers would recognize them as language. Presumably intelligent extraterrestrials would not possess human language, but would be likely to be able to pick out non-random coded signalling as evidence of intelligence, and to use them as a means of exchanging information.[42] If we believe that such an exchange between human and non-human would have great significance for our philosophical views of the world, perhaps we should be more interested in coded information exchange with other terrestrial intelligent species, like the chimpanzee.

Edward Regis argues that communication with other intelligent life can only be possible and/or significant to either party where they are sufficiently similar to have some common ground and sufficiently different to make 'comparing notes' an interesting exercise.[43] If he is right, then communication with the apes and cetaceans is arguably much more likely to be fruitful than attempts to contact aliens (if they exist) simply because we know that the necessary similarities and differences are there.

[41] See Hans Freudenthal's book *LINCOS* (Amsterdam: North Holland, 1960), for a detailed discussion of the basis for designing a 'language' (LINgua COSmica) for communication with extraterrestrial life.

[42] I think it is possible to imagine intelligent beings who would be simply baffled by language-like codes; for example, an intelligent colonial organism whose members possessed nervous connections between their brains would never need to develop any external method of signalling.

[43] Edward Regis, Jr., 'SETI [Search for ExtraTerrestrial Intelligence] Debunked', in Edward Regis, Jr. (ed.), *Extraterrestrials: Science and Alien Intelligence* (Cambridge: Cambridge University Press, 1985), 231–44.

5

ANIMALS AS PART OF
THE ENVIRONMENT

It is sometimes suggested that the problem of how we ought to behave towards other animals is merely a part of the question of our relationship with nature. In this chapter, I intend to show that this idea leads to confusion and muddled thinking. The question of animal rights is essentially based upon the significance of the subjective experience of the individual: environmental questions are much more closely related either to purely prudential problems of human survival or to questions about aesthetics and the maximization of abstract qualities such as beauty. The questions of animal rights and environmentalism may often have significant convergence with respect to the kinds of detailed plans of action we may be moved to adopt after considering them. However, this convergence is not essentially different from the kind of convergence which exists between the questions of human rights and environmental protection, since *one* reason for preserving the natural environment of Earth is concern about the bad effects of environmental degradation upon conscious beings, who may be human or animal.

Because the two issues are fundamentally separate they can sometimes conflict: the species' good is not inevitably identical to that of its individual members, and sometimes conscious animals may act in ways which alter natural ecological communities. When this happens we have to balance the different requirements of animal rights and environmental protection and it is only possible to do this if we are aware that we are dealing with two issues not one.

I am not claiming that duties to species and individuals are different *because* they conflict, but that they can conflict because they are different. One type of duty (to individuals) refers directly to the (independent) value of their mental states, while the other refers to our valuation of species as metaphysical entities. Understanding this helps us evaluate the weight we give to the two sorts of value. We may feel that respect for animals is one element of a theory of animal rights; such respect would tend to make us feel that extinction of a conscious animal species is a greater tragedy than, say, loss of a particular plant or bacterium. However, such concern clearly remains subordinate to the issue of rights. It is not the case that an 'environmental ethic' is a more basic version of a theory of animal rights.

The assumption that conservation is, or should be, identical with concerns about individual animals leads to a considerable muddle and bad temper. See, for example, complaints by RSPCA volunteer workers that the RSPB (Royal Society for the Protection of Birds) does not provide help for bird casualties.

> . . . the Newcastle upon Tyne Branch [expressed the concern] that the joint RSPCA/RSPB publication on the treatment of sick, orphaned and injured birds seemingly accepted that all practical work should be done by the RSPCA. If the RSPB, which enjoyed enormous public support, was not prepared to care for birds, despite its title, then the public should be made fully aware of this fact.
> Mr Stefan Ormrod . . . explained that . . . the RSPB's main object was the *conservation* of wild birds by the management of bird sanctuaries and reserves. The RSPB view was that financing bird hospitals would detract from this work and have little conservation value. . . . Other speakers nevertheless felt that the RSPB's limitations should be strongly publicised.[1]

Conversely, when canvassing support for the Islington Animal Charter (which states *inter alia* that animals do have rights) I discovered that many people were unable to understand why a charter for animal protection should be concerned with the feelings of individual animals, and were puzzled why domestic animals should be considered at all, since such concern was seen as not being 'ecological' (whatever that is). Even philosophers seem to feel that it should be possible to place 'animal rights', and the subject of our relationship with non-human animals, neatly in the 'environment' pigeon-hole.[2]

---

[1] Anon., 'Branch Officers' Conference', *RSPCA Today*, 52 (Summer 1986), 8.

[2] As, for example, the Society for Applied Philosophy at its 1986 Annual Conference, theme: 'The Environment'; and John Rawls, *A Theory of Justice* (Oxford: Oxford University Press, 1972), 512.

Opponents of increased consideration for individual animals sometimes make use of the need for conservation as a diversionary tactic:

Although according significant moral rights to animals appears not to have survival value to the culture, this does not imply that concern for their welfare will have no place in a highly evolved culture. On the contrary, it can be expected to be an important feature of it for at least two reasons. First, conserving the wide variety of species that inhabit our planet is important to the culture. Species are interrelated, and a drastic decrease in the numbers of one often adversely affects others, including our own. The presence of a large variety of species on our planet is also esthetically reinforcing to many people, perhaps at least partly because our species evolved in an environment that was rich in life forms. And, of course, the availability of a large variety of species to observe and study is useful for instructional and scientific purposes.[3]

I hope to demonstrate that the subject of animal rights falls most conveniently within the general topic of rights and justice, rather than that of duties to the biosphere as a whole, and that a division along these lines involves less strain on our existing moral concepts than attempts to devise an environmental theory of rights. More importantly, it makes it possible for us to articulate concern about the sufferings of conscious non-humans who are not independent of human society.

That an ethic which depends solely upon the significance of communities of organisms is insufficient to satisfy our ethical intuitions can be seen by considering the position of moral agents who have become detached from their biotic community—for example, by leaving Earth in a spaceship. Such people would still want to use ethical reasoning to decide how they ought to behave towards one another, and such reasoning could only be based upon the intrinsic value of conscious persons. Stretching imagination a little further, one could imagine a situation like that described in a short story by Eric Frank Russell, in which space explorers were accompanied by companion animals in order to safeguard their mental stability.[4] In this situation, I

[3] F. L. Marcuse and J. J. Pear, 'Ethics and Animal Experimentation: Personal Views', in J. D. Keehn (ed.), *Psychopathology in Animals* (London: Academic Press, 1979), 311.

[4] 'Hobbyist', in *Like Nothing on Earth* (London: Dobson, 1975), 7–47.

believe we can see that human and animal would be likely (as in Russell's fictional narrative) to have the kind of experience which I think is one basic component of ethical thinking: 'worrying about' one another. In the case of most non-human animals this kind of concern is unlikely to attain the degree of rational insight which would qualify them for the status of moral agents; however, the human members of such a partnership surely are true moral agents, reasoning ethically about what is due to their companions. It seems to me we are then compelled to say that this is a situation in which non-human animals do have rights (because of the companionship they provide, the human partners *want* to recognize their rights, so there is no problem about the difficulty of attributing rights to beings who are powerless to enforce them); and such rights depend upon the animals' status as conscious, social beings (plants are no good at all as sources of social companionship). But if we are prepared to admit these special cases, it no longer makes sense to claim that value can reside *only* in the totality of the biosphere. We must at least accept that there can be a separate source of value in the experiences of conscious individuals. It might be suggested that the space-voyage scenario represents an extension of the biosphere beyond earth. However, it seems to me that there is a clear difference between the case described, and one in which space travellers took with them useful food plants or inorganic materials. Secondly, I think that the idea of deep ecology has to involve the belief that natural ecosystems are valuable in a way in which ones which have been organized by humans are not. If this is not so, then there seems to be no way in which a deep ecologist could use the theory to make moral decisions: a radioactive desert is *an* environment (surroundings) just like any other. Alternatively, if the deep ecologist wants to argue that the value of any ecosystem, whether natural or artificial, depends upon the nature of characteristics which it possesses, such as beauty, or stability, then it seems to me that the existence of conscious beings in the system must be one element in its evaluation.

Deep ecologists are free to claim that, where there is conflict between the two sources of value, environmental value always trumps that of states of consciousness. Even so, value from consciousness could still be the deciding factor in some important areas of decision-making: for example, if environmental concerns dictate that a population of animals should be controlled, respect for individual rights would still have an important role in determining what kinds of methods might justifiably be employed. Animals which never form parts of natural ecosystems (such as experimental guinea-pigs) are essentially in the same position as the fictional ones in the spaceship. It seems to me that attempting to pigeon-hole our concern about them as concern about the environment simply strains the concept too far, while it appears quite reasonable to say that we are concerned about them because they are conscious creatures who may be exposed to unnecessary suffering or distress.

Some authors[5] have claimed that making a distinction between the moral status of conscious and non-conscious members of the biosphere (such as a distinction between animals and plants) involves mere anthropocentric prejudice. I should like to suggest, on the contrary, that there is a fundamental and objectively verifiable difference between those animals which seem plausible candidates for consciousness, and all other non-human members of the biosphere: the capacity for choice. The question whether a particular tree 'wants' to continue standing is meaningless, but we can often find out what animals prefer, given a choice of several alternatives. This knowledge seems to impose constraints upon our actions of a type which appreciation of the intrinsic (or inherent) value of a non-conscious being does not. If, for example, we know that pigs very much prefer to have some periods of light, rather than being kept permanently in semi-darkness (because they will operate a switch to turn on the light given the choice), this would seem to impose some sort of obligation upon humans who choose to keep pigs, of a kind which would make no sense applied, for example, to the decision whether or not to prune fruit trees, which can have no preferences one way or the other. Pigs appear far more readily comparable to human workers, in respect of whom it is natural to say that it is simply unfair that their conditions should be

[5] e.g. John Rodman in 'The Liberation of Nature', *Inquiry*, 20 (1977), 83–145.

organized with no thought at all to their own preferences except in so far as these directly affect their capacity to serve their employer. The question of objective measurement of what animals prefer is reviewed in Marion Stamp Dawkins's book *Animal Suffering: The Science of Animal Welfare*.[6] She makes the sensible point that we do not always feel we should respect animals' preferences (e.g. not to go to the vet). However, I suggest it is really not at all difficult to construct a common-sense attitude towards knowledge of what animals choose, and that this attitude need not differ radically from our beliefs about human choices. Most of us (I think) assume that we may override human autonomy if someone is clearly in extreme danger of self-destruction (such as committing suicide). This assumption seems to be grounded in the presumption that a person *would* choose to be saved if she had normal command of her faculties, or if she had full knowledge of the situation. It is perhaps helpful to make a comparison with attempts to draw up sensible general principles about the rights of children. C. A. Wringe, for example, makes the important point that it is quite possible for a child to be capable of making rational decisions about some kinds of problem and not others, and that it makes good sense to say that children should have rights of freedom in respect of the former even though it may be quite proper for adults to coerce them in situations where they are not capable of understanding their own best interests.[7] (Wringe uses the example of a 4-year-old child's ability to decide whether he wants to play with his toy train or his scooter and his inability to make a rational choice whether to have a polio vaccination or not.)[8] In the same way we can sensibly say that, for example, farm animals have rights of freedom to be allowed to turn round, lie down, groom, stretch, and so on, but must be subjected to some kinds of restriction, both for their own benefit and for the protection of other conscious beings. In general, the desire of a cow to turn round or stretch to make herself more comfortable is likely to be just as rational as our own desires to do these acts. On the

[6] (London: Chapman and Hall, 1980). ch. 7.
[7] *Children's Rights: A Philosophical Study* (London: Routledge & Kegan Paul, 1981).
[8] Ibid., p. 111.

other hand, cows can't be expected to understand that they are liable to cause accidents to themselves and to others if they are allowed to wander freely along busy roads. One might even reasonably agree with Wringe that there can be a welfare right to receive appropriate coercion when this is necessary to preserve an imperfectly rational being from the effects of misguided choices.[9] Thus, one might say that a person who fails to ensure that her dog is vaccinated against distemper, because the dog dislikes the process of vaccination, is offending against the dog's welfare right to protection. Similarly, allowing a dog to roam freely and attack sheep offends against the welfare right of the sheep to security. (This need not mean we have a duty to restrain truly wild animals from their natural actions as predators: as members of human social groups domestic animals like dogs stand in a special kind of relationship with us.)[10]

The preservation of species may well not coincide directly with the interests of individuals, and must be seen as a completely separate question. I do not, however, think that species preservation can be entirely ignored when we ask questions about our duties to animals: the simple assumption that 'they would be better off dead' is surely as arrogant in its own way as the 'postage-stamp' mentality which sees the preservation of a few 'type' specimens in zoos as an adequate discharge of our duties to (or about) wild animals.

In all of the furore about preservation it is seldom questioned what species are to be preserved *for*. If the answer is in order that possible damage to the life-sustaining systems of the biosphere may be averted, then we should not be concentrating upon animal life at all, certainly not upon the preservation of the larger, rarer, and more spectacular mammals. The protection of plants and micro-organisms, which normally engages our attention to a much lesser degree, ought then to be the priority. Indeed, animals who are preserved only in zoos are not part of the natural ecosystem at all.

As pointed out by Jeremy Cherfas, preservation of animals in zoos depends on maintaining sufficient genetic diversity to avoid the deleterious effects of inbreeding on animals' health

[9] Ibid., p. 138.
[10] S. R. L. Clark, 'The Rights of Wild Things', *Inquiry*, 22 (1979), 171–88.

and fertility.[11] Since there is a limit to the number of animals zoos can afford to feed and house the rate at which animals die must balance that at which they are born. However, nearly all animals will breed much faster than the rate at which they die (or have to be euthanized to prevent suffering). It is possible to restrict the birth-rate by contraceptive methods, or by neutering animals once they have replaced themselves in the captive population, but this means that either the effective genetic population will be only a small fraction of the total number of animals kept (since for each female who is breeding for the first time the zoo must maintain her mother, grand-mother, great-grandmother, and so on, in a non-reproductive condition), *or* animals must be prevented from breeding until relatively late in life. The second alternative poses risks for the species because fertility tends to decline with age and it would be fairly easy to drift imperceptibly into a situation in which most species members preserved in zoos could no longer contribute to the next generation.

The alternative to control of the birth-rate in zoos is an increase in the rate at which animals die. By killing animals once they have provided their own replacements the total breeding population (and hence the genetic diversity) can be maximized, and this represents the surest way of preserving the species. This solution is the one favoured by Cherfas, who considers that the welfare of individual animals should be subordinate to that of the species to which they belong. He believes that the present situation, in which zoos which kill 'surplus' animals are subjected to 'muddled public outcry' (p. 122), is the result of lack of education of the zoo-going public. Apart from the question of whether it is really possible to equate the welfare of species (which cannot possess any personal experiences) with that of individuals, who can, it is perhaps questionable whether the reaction of zoo visitors to the killing of healthy animals is really the result of lack of education. Cherfas on the whole views zoo animals primarily as a scientific resource, although he is sensitive to the need to prevent suffering, and clearly actually *likes* wild animals. However, it does not seem to me that other people who visit

---

[11] *Zoo 2000* (London: BBC, 1984), 108–22.

zoos (and incidentally who pay for their continuance) can be required to adopt his viewpoint, rather than, for example, seeing zoo animals as a type of companion animal.

If they happen to disagree with Cherfas, it is not necessarily the case that education would, or should, change their views. It might: clearly the strictly best option from a conservation point of view is not compatible with a view of zoo animals as companion animals whose individual life and welfare is a primary aim, and a zoo-goer who presently believes that she is primarily in favour of zoos because they conserve animal species, but who also opposes the killing of healthy animals, must revise her opinion either about the importance of conservation or about the value of individual animal lives. However, there is no clear reason why a zoo visitor should not decide that conservation should be subordinate to the rights of individual animals, rather than the other way round.

If animals are maintained in zoos as a last resort, we have a choice whether to allow them to adapt to conditions there. The animals best suited to zoo life will tend to produce more offspring than those less suited and the population will drift away from the wild type unless artificial methods are used to ensure that all animals contribute equally to the next generation. If there is virtually no chance that ancient habitats will be re-created in the future, the interests of individual animals weigh on the side of permitting a degree of adaptation and integration into human-controlled surroundings, which of course is domestication. But is this any more 'artificial' than forcing zoo animals to remain permanently maladapted to the environment they actually have to cope with?

In addition to a duty to attend to rights, it may well be true that we ought to respect the marvellous complexity of inter-action between all kinds of life, both conscious and unconscious. Smart uses a similar argument in favour of our duties to respect other human beings, drawing an analogy with our intuitive feeling that wanton destruction of some intricate piece of human workmanship, such as a watch, is intrinsically wrong. He remarks:

A man who has any feeling at all for machinery will be extremely reluctant to destroy even an aeroplane or a clock. And yet there are men who think nothing of burying an axe or a bullet in the beautifully organised brain of a fellow-creature.[12]

We may also feel intuitively that, for example, the oldest living thing in Britain (an ash tree which is estimated at over 1,000 years)[13] has a significance which is independent of the capacity for consciousness. Certainly the wanton destruction of such an object seems to indicate an important lack in the human personality.

There is at present a considerable degree of controversy over the extent to which life on Earth can be held to form a harmonious whole (the so-called 'Gaia' hypothesis)[14] as opposed to a merely opportunistic association of organisms struggling for their own survival. The two ideas may not be entirely incompatible. Genetic theory predicts that organisms cannot diminish their own genetic success for the benefit of the larger biotic community. However, we have some evidence that, on at least two occasions in the past, independent genetic lines came together to form associations. It is thought that the ancestors of modern plants and animals were formed by symbiotic association of simple, bacteria-like organisms belonging to several different species, which 'pooled' their various specializations. Thus, one bacteria would contribute its efficient energy-mobilizing capacity; another contractile proteins for mobility; yet a third the capacity to use light for photosynthesis of carbohydrates. Presumably this resulted in net genetic benefit for all the partners. These complex symbiotically derived cells are collectively termed eukaryotes, while simple, bacterial cells are known to biologists as the prokaryotes.[15] The mitochondria, which are the 'power-stations' of our body cells, have their own DNA, which is independent of the cell nucleus. Similarly, the chloroplasts of higher plants have their own DNA, replicate by division, and cannot be

---

[12] J. J. C. Smart, *Philosophy and Scientific Realism* (London: Routledge & Kegan Paul, 1963), 152.

[13] O. Rackham, *Trees and Woodland in the British Landscape* (London: Dent, 1976), 29.

[14] J. E. Lovelock, *Gaia: A New Look at Life on Earth* (Oxford: Oxford University Press, 1979).

[15] For a detailed discussion of the symbiotic theory of the evolution of eukaryotic (chromosomally organized) cells see L. Margulis, *Symbiosis in Cell Evolution* (San Francisco: Freeman, 1981).

made by nuclear DNA acting alone. It is thought that both kinds of organelles are derived from symbiotic bacteria. Such components submerge their independent interests in favour of the success of the whole, presumably with net genetic benefit for all of them. The growth rates of the symbiotic partners are regulated so that symbiotic organelles neither destroy the host nucleo-cytoplasm by overgrowth, nor become too scarce to perform their functions in the partnership. Symbioses which are not at present so firmly obligate as those between the different parts of eukaryotic cells (for example, those between termites and the various protozoan symbionts which enable them to make use of dry wood as food) also show adaptations which tend to maintain the partnership in a balanced form. Large termite protozoa are closely associated with motile bacteria (spirochaetes), which attach to the protozoan's surface and help to prevent the protozoa being lost when the termite defecates. When the termite moults it sheds the lining of its hind-gut along with its external skeleton, and the symbiotic organisms are temporarily lost. However, the termite's moult hormone also acts as a signal for its symbionts to form dormant spores, which are returned to termite guts when the moulted skeleton is eaten.[16]

It is not totally implausible that similar meshing of interests could have operated to generate ecological control systems which operate above the species level (although still based on the 'selfish' gene). The symbiotic systems which make up multicellular organisms could perhaps be seen as microcosms of a larger, Gaian symbiotic system. And, indeed, soil scientists have found it useful, for some purposes, to treat terrestrial soils as if they were tissues or organs in which certain functions are carried out by different groups of organisms: decomposers, nitrifiers, denitrifiers, and so on.[17] Thus McCloskey is mistaken in his belief that truly interesting ecological communities are to be found only in remote places accessible only to an affluent élite.[18] Any gardener is reliant upon complex biological cycling

[16] Margulis, *Symbiosis in Cell Evolution*, pp. 172–94.

[17] A. D. McLaren and George H. Peterson, 'Introduction to the Biochemistry of Terrestrial Soils', in A. D. McLaren and George H. Peterson, (eds.), *Soil Biochemistry* (London: Edward Arnold, 1967), 1–14.

[18] H. J. McCloskey, *Ecological Ethics and Politics* (Totowa, NJ: Rowman & Littlefield, 1983), 42 f.

of nutrient chemicals by associations of wild organisms which are certainly not yet fully understood, much less controlled by humans.

However, there is one enormously significant difference between evolution of symbiotic organisms and the process which yielded the present-day biosphere of the Earth. A partnership of several bacteria can replicate to produce more partnerships, and it will be more or less successful in doing so than other competing partnerships and individual cells. Gaia is not a replicating association, in the sense of being capable of generating 'daughters', although it is possible for the biosphere's total mass to increase or decrease. Hence (for example) an over-successful organism which is liable to overgrow the whole can only be checked by the tendency of other members of the biosphere to evolve defences against its competing effects. Where such over-success happens in a variant of a population of a symbiotic species, evolution proceeds more directly: the variant is unsuccessful in comparison with balanced symbioses, which replicate while it declines. Thus, it is clear that it is in the genetic interest of symbiotic partners to restrict their own growth to maintain balance, while it tends to be in the interest of the members of Gaia to evade restrictions. (The problem is rather similar to that of species versus individual success in the mountain gorilla discussed in Chapter 9: short-term genetic success always has a tendency to trump long-term success.) Gaia's only possible way of evolving lies in the genetic 'interest' which its members have in preventing all other members from becoming too successful.

It is perhaps significant that the most convincing proponents of the value of ecosystems (at least to me) are those like Margulis and Lovelock who clearly find them fascinating while still retaining a healthy respect for the value of individual creatures. This contrasts with authors such as John Rodman who seem at times primarily motivated by the belief that humanity corrupts anything it touches.

The rights of conscious beings could sometimes conflict with 'letting nature take its course' without necessarily meaning the disruption of ecosystems. For example, suppose that a population of animals is at present regulated by periodic outbreaks of a painful disease, and that it would be possible for humans to control the disease and replace its controlling effect by contraceptive methods. This is not a purely fanciful suggestion: vaccination of foxes has been used in attempts to control the spread of the rabies virus[19] and neutering campaigns have been used to control feral-cat numbers. A combination of the two is not an unreasonable idea. This kind of alteration in a natural system would clearly be a beneficial one on a utilitarian calculation of interests, based upon consciousness. It seems to me that, if deep ecologists like Rodman wish to argue for the overriding importance of entirely natural systems, the onus of proof must lie with them.

Purely 'ecological' theories of value, such as that proposed by Rodman, do not admit that we can be supposed to have duties towards domesticated animals (except perhaps the duty to exterminate them as rapidly as possible). I believe that this attitude can be shown to be a mistake even on the supposition that only ecosystems have value, since it is a fact that domestic animals play essential roles in many existing 'wild' habitats. For example, the British Isles contain wetlands which are important wintering areas for large proportions of the European wildfowl populations, including the endangered Greenland whitefront goose and Bewick's swans. The suitability of these wetlands for birds depends crucially upon grazing by domestic stock during the summer months, since ungrazed wetland pasture will tend to revert to damp willow and alder woodland. Rodman could perhaps claim that we are obliged to try to re-create self-sustaining ecosystems which are entirely independent of human activity. If so, I think he has to show that this has some possibility of success: it is not likely, for example, that the inhabitants of East Anglia could be persuaded that reflooding the Fens would be desirable. He must also supply some answer to the question 'To what point in time must we turn the clock back?' Human activity has been modifying ecosystems in Britain at least since the last glaciation and there may well be *no* natural ecosystems which exist here if, by natural, we mean ones which have never been affected by humans or domestic animals. I don't think we can be supposed to have an obligation to try to produce ecosystems

[19] D. W. Macdonald, *Rabies and Wildlife* (Oxford: Oxford University Press, 1980), 114–37.

which have never existed before but which might have done if *Homo sapiens* had never evolved.

Furthermore, the majority of land-based wildlife reserves now depend upon human management to control the balance of species because they are too small for natural regulation processes to work. An example of this problem is that of the regulation of numbers of elephant in Africa, where these animals have fled from persecution by hunters and now threaten their own survival by over-grazing the relatively tiny areas of land allocated for them. Here, consideration of interests of conscious beings must have some part to play in determining what methods of control we may justifiably use.

Deep ecologists who take Rodman's view of domestication must also explain how this process of adaptation to human society differs in terms of value from changes in animals who associate with humans while remaining undoubtedly wild, such as garden birds. Bottle-opening by blue-tits is clearly an 'artificial' activity; it most probably involves an element of conscious choice; it certainly is not a human activity. Why is it only *human* chosen activities which can be 'unnatural' in a bad sense? If, on the other hand, bottle-opening by birds is a bad thing (because it is unnatural), none of the animals who relies to any extent on learned behaviour can be said to be truly natural. This is clearly silly, on a par with the value judgements of some bird-watchers who determine whether mute swans are 'worth seeing' or not on the basis of current ideas about the latter's status as an indigenous species (good) or naturalized species (bad).

I suggest that it makes more sense to say that human activities are a 'bad thing' where they cause suffering or loss of capacity in the animals affected, because those animals have interests in avoiding suffering and enjoying conscious experience. On these terms, for example, selection of jaw malformations in Pekinese dogs is bad, selection for greater sociability in cats is probably on balance good. Injuries and harms need not necessarily involve suffering to be a cause for moral concern. Loss of capacity (e.g. making animals less intelligent by keeping them in featureless surroundings) would count as a morally significant injury. But, I think, there is a clear distinction between this and damage which has *no* effect on any consciousness (for example, causing physical infertility in a woman who had no intention of having children anyway). It seems to me that damage to a non-valued plant falls into the second category and hence need not normally be a reason for concern.

I would further note that many domestic animals have more control over their own lives and genes that we sometimes suppose, and adaptation to human society does not necessarily mean that humans have made conscious decisions to mould animals to their design. Some degree of mate choice still operates except where artificial insemination is the only method of breeding allowed, and there may be a surprising amount of interchange between domestic and wild or feral populations. Where a genetic change is selected because humans are unconsciously attracted by the phenotype produced (for example, juvenile behaviour) it is arguable perhaps who is exploiting whom. If such domesticated-animal behaviour is to be stigmatized as degenerate[20] it seems to me that Rodman must commit himself to saying that most symbiotic organisms living in non-human systems are also degenerate. But if he permits himself to criticize 'nature' to this extent, why not go the whole hog and agree with Singer (and Darwin!) that there are other aspects which are not desirable, namely, the amount of suffering which takes place in nature. It is, after all, true that all conscious creatures try to avoid pain and suffering and Rodman's pervading sense of corrupt humanity viewing 'good' nature from the outside seems to me to involve more genuine alienation from the natural state than Singer's perception of us as animals together. Pet-keeping *can* involve unacceptable lack of attention to animals' behavioural needs. Bonnie Beaver quotes a case in which a kitten was given anti-epileptic treatment in attempts to control her normal play behaviour because her owners believed that her sudden 'attacks' of running about were some kind of fit.[21] However, people who enjoy the company of domestic animals are not usually so unperceptive as this.

As I have indicated above, I think that concern about our duties to individual animals and concern for the totality of the

[20] Rodman, 'The Liberation of Nature', pp. 83–145.
[21] *Veterinary Aspects of Feline Behavior* (St Louis, Mo.: Mosby, 1980) 42.

biosphere are two respectable, but distinct enterprises. However, interest in the value of ecosystems could be usefully integrated into a complete theory of our duties to animals to explain why there is a difference between our duties to animals who live as members of human society and duties to animals who do not. It seems to me that there is a plausible analogy with our different duties towards human members of our own society and those belonging to other groups. Our predominant duty to other groups is to leave them in peace to arrange their own affairs, although in extreme situations we may have duties to respond to requests for assistance. Members of a single social group have additional rights to help and protection; but those responsible for ordering the affairs of the society (who may be as many as all competent adults) have rights to control the activities of society members in the common interest. Society members who are also moral agents have duties to comply with reasonable rules for the ordering of society. This would explain why we have duties to prevent our dogs from worrying sheep, and should interfere when cats try to capture birds, but we ought not to interfere with predators in natural ecosystems who are hunting other animals for food.

Reasoning about animals' choices by analogy with our own feelings probably does sometimes involve us in mistakes, but the belief that all members of natural communities enjoy a calm sense of their overall place in the food chain[22] clearly is a total fantasy. I don't claim to know exactly what a wood mouse feels when she is being crushed by an owl's talons, but I am quite certain that 'pain' and 'fear' would express it more closely than Bellamy's peaceful meditation on the flow of nutrients in an ecosystem.

I knew that my life was about to come to its end—and yet rejoiced in the fact that some of my substance would become part of the Owl or her new brood and know the freedom of flight. I also knew that my least tractable remains, teeth, bone and fur would soon be pelletised upon the forest floor and would provide food upon which fungi and bacteria would grow, releasing the elements that had made me what I am . . . But what of all my experiences, my loves, my hates, my mind so packed with memories past and questions brimming about the

[22] As, for example, in David Bellamy's charming book *The Mouse Book* (Stocksfield, Northumberland: Oriel Press, 1983), 59.

future, where would they go? . . . I turned in vain to see what species of bird I was about to become—and gloried once again in three hundred and sixty degrees of firmament . . .

This kind of thing is, I think, the main danger of putting too much of a premium upon 'environmental ethics'. Respect for the choices of others does help to keep us in touch with reality.

The concerns of 'shallow' environmentalists might often be more satisfactorily addressed by appeals to the concept of rights than by emphasis on the idea of 'environment'. For example, if a particular factory is being run in a negligent way which threatens harm to the safety or health of people who live near by, I am not clear what is added to statements about the wrongness of the factory managers' actions by claiming that this is an *environmental* problem. How does this wrongness differ significantly (for example) from the wrongness involved if I were to endanger people by negligently continuing to drive a car which I knew to have defective brakes, and why would it not be sufficient to say that in both cases people have a right not to be endangered by negligent actions? I suspect that, here appeal to 'environmental' principles is (at least tactically) weaker than use of the rights argument. Consider, for example, these two possible replies to a complaint about poisonous emissions from a chemical plant:

To the environmental complaint the answer might be, 'Of course protecting ecology is important—but wealth creation and jobs have to come first.'

But a similar reply to the rights complaint sounds very much less plausible: 'Of course human rights to freedom from injury are important—but wealth creation and jobs have to come first.

Again, if we are trying to argue that a group of South American Indians ought not to be dispossessed from their land and made destitute, what exactly does the statement that this is an environmental issue add to the claim that the Indians have a *right* to their own territory, and ought to be treated like any other nation of human beings? My point is that it is easier to portray concern about something as undefined and vague as 'ecology' or 'environment' as something which 'realists' have to subordinate to practical concerns (unlike sentimental idealists). It is not so easy to shrug off the protest that real, identifiable individuals are having their rights infringed if they are injured by other people's money-making activities. I believe questions about the rights of conscious animals almost exactly parallel these kinds of human examples and that we should avoid getting into a position where we feel we must try to 'fudge' ecological arguments when we are really trying to express concern for the welfare of conscious individuals. On the other hand, I would like to suggest that it is helpful to see the promotion of interest in ecosystems, over and above what is necessary for the survival of conscious entities, in terms of education of taste. This clearly can be morally significant if it enlarges the capacities of persons for experience and enjoyment.

From a purely practical point of view, it is necessary that people who are interested in protecting the welfare of individual animals should take account of natural ecological laws. However, this is no different from the requirement that people engaged in practical reasoning about human affairs should take account of the laws of nature.

6

KILLING ANIMALS

The issue of trying to minimize pain and suffering in animals poses a major paradox. If we admit that pain is undesirable because pain hurts, then to be logically consistent we would have to concede that it is alright to kill animals (as long as the technique is painless), since most species are oblivious to their own existence and therefore cannot concern themselves with their eventual demise. Yet, the question of whether an animal lives or dies seems to me to have consequences which are far more profound than whether it experiences an occasional aversive stimulus. . . . in mindless species the experience of pain may be different than it is for you and I. . . . Pain, hunger, fear, frustration, and thirst serve to activate behavior sequences which have hard-wired adaptive significance, but lacking the ability to monitor such states they may be unconscious.

Chimpanzees, orangutans, and maybe cetaceans and elephants pose a fundamentally different ethical dilemma. . . . Can we continue to justify incarcerating, killing, and abusing such life forms? When a mind dies a universe blinks out of existence.

Gallup, in *Animal Cognition and Behavior.*

The confusion between consciousness and self-consciousness illustrated in this quotation has been considered in Chapter 3. However, the questions raised about the relative significance of suffering and death and whether this is the same for all species of conscious animals are important ones. What bearing has evidence about the existence and nature of animal consciousness upon what constitutes a harm for members of different species? The sensations experienced by animals are sources of value, good or bad, just as our pleasant and unpleasant sensations are. If we believe that moral systems ought to take account of value, then any satisfactory moral system has to give some account of the way we ought to treat non-human animals as well as a theory about relationships between humans.

But animals are subjects, not objects, and as such are entities with a particular individual value, independent of their status as 'containers' of utility. This has particular bearing upon the morality of killing animals, since it seems sometimes to be thought that, while there are moral problems about causing animals to suffer, we need not worry about ending animal lives, provided that we do it painlessly. It might be claimed that non-conscious material objects are as valuable as conscious individuals. I suspect that this claim depends at least partly upon a limitation of the imagination, specifically the inability to comprehend the possibility of the extinction of one's own consciousness. The proponent of the value of material things (at least so it seems to me) has a mental picture of trees, planets, stars, and so on continuing their existence in majestic silence after the extinction of the last conscious observer. But, without conscious minds there *is* no picture: merely blankness. Even the most humble consciousness (say, that of a rabbit) produces a situation which is radically different from a non-conscious universe. Even the most perfectly adapted mindless organisms can generate nothing like this difference in kind.

Utilitarians attempt to explain why we normally feel that killing is a wrong (at least in the case of self-conscious humans) by pointing out that if a conscious being is killed this means the elimination of future chances of pleasant experiences and also that it normally means other individuals suffer the experience of fearing that the same fate will befall them. If animals who would otherwise have had pleasant experiences are killed and not replaced this means that the total utility is diminished, so there is at least some degree of wrong involved even when those killed are perhaps only minimally self-conscious. It is difficult to describe this killing as a harm or deprivation to the particular animals who are killed, since no conscious being is left in a deprived or frustrated state and the event is said to be of moral significance only because of its effect on the sum of the utilitarian calculus. This is one reason why classical utilitarian systems have difficulty in explaining our intuitive feelings about the wrongness of killing in terms of harm to the individuals who die. Preference utilitarianism attempts to incorporate this intuition by claiming that the principle of utility should mean that we should maximize fulfilment of the preferences of conscious beings; hence, if a being has an overall preference for remaining alive, killing that being is a harm because it is counter to that preference.

Bernard Williams offers a somewhat similar suggestion that the evil of death lies in the frustration of what he calls categorical desires, that is, desires which do not (like the desire to avoid pain) exist only on the assumption that one is going to be alive.[1] He suggests that, once a person ceases to have any categorical desires death is no longer an evil for that person, and that it is not possible to have the desire to live in the absence of other desires.

These theories, and variants of them, suggest that simple, non-reflective consciousness is not sufficient to make killing a harm. Animals who are simply conscious, or have minimal self-consciousness, will not be aware of themselves as beings which exist over time, and hence will not be capable of desiring to continue to exist over time. However, this leads to some problems in explaining (for example) why it is wrong to kill humans who temporarily lack a preference for continued existence, such as persons who are asleep. In practical terms it may not raise difficulties for our intuitive belief that killing must be a prima-facie harm. Animals with only limited self-consciousness may still be frightened by the sights and noises of attempts to catch or kill them. For example, it is evident that wildfowl are afraid of the noise of shooting. It is probable that many prey species have an innate ability to recognize when potential predators are hunting. Behaviour patterns such as staring and stealthy, stalking movements tend to be common to most predators, and even humans trying to catch farm animals may act in a very similar way. Where this happens, the fear experienced by the animals is a sufficient reason for us to say that they are indeed suffering a harm, although it might be claimed that, for example, leaving them to multiply until their numbers were checked by starvation would be a still greater one. Thus, even if we believe that

[1] 'The Makropulos Case: Reflections on the Tedium of Immortality', in *The Problems of Self* (Cambridge: Cambridge University Press, 1973).

killing is not a harm to non-self-conscious animals unless
there is contingent suffering caused by the process of killing, it
may in practice not be possible to eliminate such suffering to a
point at which killing does not represent harm.

I suggest that simple consciousness is sufficient to make
killing a harm even when there is no attendant suffering for
two reasons. Firstly, even if a particular animal is not able to
understand the meaning of existing over time it remains true
that she does so exist and that we are able to see that killing
her puts an end to her future enjoyment just as it would to that
of a human. It is not necessary to possess the concept of a
particular good for it to be possible to be deprived of that good.
Secondly, I think that the frustration of preferences for life is
not really a very satisfactory explanation of the wrongness of
killing humans. After all, if frustration of preferences were the
only reason why dying is undesirable, why should we try to
avoid death as much as we do?

The problem is rather similar to that involved in the famous
Epicurean assertion that fear of death is irrational because after
the annihilation of consciousness we shall not be around to
experience anything. This assertion and the related utilitarian
claim that death is mainly an evil because of the fear it causes
to self-conscious beings (and thus is not really an evil for
beings who do not understand the concept of death) generate a
similar sense of dissatisfaction about the way they simply
refuse to accept the widespread human conviction that death
actually is an evil in itself. Some people might query whether
this conviction really exists, but see, for example, Michael
Slote's paper 'Existentialism and the Fear of Dying':

. . . I have assumed that people who live objectively and say that they
are not terribly anxious about dying are nonetheless afraid of dying at
some level. And this may seem high-handed. However, I am inclined
to think that in general people living world-historically (who do not
believe in some traditional type of life after death) continue to be
subject to a certain welling-up of death anxiety that can overtake
them in the midst of their daily lives. Despite my own tendencies to
the world-historical, I have often experienced this sudden welling-up
of death anxiety . . .[2]

[2] Michael A. Slote, 'Existentialism and the Fear of Dying', in J. Donnelly
(ed.), *Language, Metaphysics and Death* (New York: Fordham University
Press, 1978), 73.

The idea of something which is innocuous in itself and only becomes evil when we think about it seems a rather peculiar one.[3] It appears more in line with our intuitions to say that the annihilation of a consciousness is itself dreadful and that self-conscious fear of death is a superadded evil in the same way that frightening or painful treatment at the time of death is an added evil for creatures with minimal self-consciousness. A comparison with our feelings about the death of very young children is perhaps useful. The evil of such deaths cannot be grounded in the children's self-conscious fears. An Epicurean might perhaps claim that it stems from the distress experienced by the children's parents. However, this seems an unsatisfactory explanation, since the parents themselves would certainly say that their distress was caused by the harm suffered by the child.

The question of whether lack of a highly developed consciousness of self significantly reduces the harm involved in killing is important for a wide range of decisions we make about the way we ought to act towards animals. For example, animal welfare societies commonly attempt to persuade people to control the numbers of pet animals by neutering females, rather than by killing unwanted young. It might, however, be claimed that, because cats and dogs probably have relatively little self-consciousness, painless killing is a less serious harm for them than an operation like neutering which involves transient suffering, even though this may later be compensated for by a pleasant life. It could be claimed that such compensation is not possible for non-self-conscious beings because there is, in a sense, no developed self which can enjoy the compensation. The objection also raises other serious questions. If it is correct, then it must also be the case that we ought to choose to kill non-self-conscious beings rather than subject them to *any* treatment which causes transient suffering (for example, setting a broken limb).

I think it is possible to demonstrate that this is misguided by considering treatment which involves some annoyance to the animal, but which could not reasonably be said to cause suffering, for example, administering tablets to treat cat flu.

[3] T. Nagel, 'Death', in *Mortal Questions* (Cambridge: Cambridge University Press, 1979).

Presumably it is reasonable to say that the cat does not have a rational picture of herself swallowing the tablets, or of herself later finding that she can breathe with less difficulty. However, it is still true that there is a single cat who experiences these things, and who benefits from the treatment. After painless killing there is no cat self to experience any possible pleasant or unpleasant sensations, but this is equally true in the case of painless killing of humans. It probably is true that humans can comfort themselves to some extent with the thought that transiently painful events will soon be over, in a way which other animals cannot, but I doubt whether we really feel that this capacity is the primary reason why we ought to treat someone with bad toothache instead of shooting her. We may, for example, transiently be so seasick that we say we just want to die quietly, but no one would seriously suggest that this temporary loss of the self-conscious desire to go on living would make it justifiable to kill humans. Similarly, it is probably true that small children tend to live in the moment and that transiently painful events completely fill their consciousness, in a way which does not happen in adults unless the pain is very severe indeed. This is probably a reason why it is wrong to employ very painful methods in attempts to extend small children's lives for relatively short periods, but it certainly is not a reason for withholding painful treatment which would save a child's life. It seems reasonable to say that killing conscious, but minimally self-conscious animals is wrong because selves are destroyed even though those selves may not have an integrated view of their existence through time.

If these kinds of reasons are accepted, it becomes clear that they must apply equally to killing of other species of conscious animals. Thus, for example, there is an inconsistency in the attitudes of animal welfare charities which advocate neutering as a more humane way of dealing with the problem of unwanted pets, while refusing to advocate that their members should become vegetarians.

It may be possible to suggest additional reasons for thinking that simple consciousness is sufficient for extinction of that consciousness to be an evil by considering some 'thought experiments'. In one of Stanislaw Lem's science fiction collections there is a story in which a scientist constructs an exact duplicate of a person, including all of his current memories, and claims that the immediate killing of the original would not mean a death, since consciousness continues in the duplicate. It seems to me (and presumably to Lem) that this does not represent a good reason why the original person should stop fearing death. After all, we do not expect the parents of twin babies to think that it does not matter if one of them dies, because at that stage the two are likely to possess virtually identical consciousnesses. Conversely, it is possible to imagine a different SF scenario, in which a person is kept alive and conscious by replacing the natural wastage of brain cells during ageing with fresh, cultured cells which are 'wired up' by the natural activity of the originals. In this second case we can imagine that the whole person could be eventually replaced by new cells, and that her mind would gradually change over time as she had fresh experiences, but that continuity of consciousness would be preserved.[4]

It seems to me that we would tend to say that the first type of preservation was no good to us, but that the second did preserve life in a valuable way, in spite of the fact that the final consciousness was not the same as the original. I suggest that it is continuity of consciousness which is the crucial element, and that this can be possessed by a being with simple consciousness even if she is not capable of understanding this continuity. Learning is a common characteristic of all the animals which seem at all plausible as candidates for possession of consciousness and it must therefore be true that they also possess continuity of consciousness. Killing is a harm to them independent of the extent to which it is liable to cause fear or pain because it ends this continuity.

I suspect that the major significance of developed self-consciousness is that it is a necessary condition of moral agency. It does not seem to me that it is possible to make genuinely moral choices without being able to think about

[4] Human nerve-cells do not divide and are not replaced in adults, so it is not true (as is sometimes suggested) that this kind of transformation happens in normal circumstances, although it is true that the cells will change their connections with time, and that there is some turnover in their chemical constituents.

one's own interests and those of another individual. Only if a being has a degree of self-consciousness does there seem to exist the possibility that she might decide to forgo her own interest in favour of another person. Even a policy of complete devotion to self-interest requires a capacity to comprehend that others also have interests if it is to be a policy and not simply a tendency to respond to things which attract or repel. Someone who could not think about her own interests would not be able to think about the possibility of going against them for the sake of something more significant, whether a principle, or the interests of another. Jane Goodall points out:

In order to be cruel, one must have the capability (1) to understand that, for example, the detaching of an arm from a living creature will cause pain and (2) to empathize with the victim. It is because we humans unquestionably have these abilities that we are able to be cruel. . . . If a group of humans behaved in the same way as the gangs of Kasekela males when they attacked their Kahama victims, the behavior would be described as cruel; so would the slow killing of large prey animals. Of course, chimpanzees are intellectually incapable of creating the horrifying tortures that human ingenuity has devised for the deliberate infliction of suffering. Nevertheless, they are capable to some extent of imputing desires and feelings to others . . . and they are almost certainly capable of feelings akin to sympathy. The Premacks' Sarah consistently chose photographs of her 'enemy' strewn with cement blocks, suggesting that she may possess some precursor of sadism. On the other hand, her motivation may have been nothing more than mischief making.[5]

Animals who probably lack developed self-consciousness do sometimes give up benefits to others (for example, sharing food). This is probably better explained by saying that they do this because of friendly feelings towards those who benefit, rather than through a deliberate intention to give up their own interest for the sake of another. (After all, even humans do not manage this very often.) It seems reasonable to claim that the annihilation of a moral agent is worse than annihilation of an agent who lacks the capacity for moral reasoning, but not that the latter is of no consequence at all.

[5] *The Chimpanzees of Gombe* (Cambridge, Mass.: The Belknap Press of Harvard University Press, 1986), 533.

In fact many birds and mammals may well be sufficiently developed mentally for some of the additional reasons why death is an evil to humans to apply to them as well. A fairly simple capacity for anticipation of future events and possession of future goals implies the applicability of utilitarian arguments about the wrongness of killing based on frustration of desires and preferences.[6] These goals need not necessarily be highly complex or sophisticated. A rat may perhaps anticipate feeding time, a dog look forward to walk, a child long for the school holidays, and so on. Thus beings capable of anticipation could be said to lose something when they are painlessly killed in a way in which ones who live purely in the moment could not. This capacity seems not to require such a sophisticated level of self-consciousness as a true preference for life. It is possible to have genuinely desired goals in life without having a genuine understanding of one's self as a continuing consciousness which will, at some time, come to an end. It may well also be true that repeated painful events are a much greater wrong to creatures with at least minimal self-consciousness, since fear of coming events might make the pain-free periods almost as miserable as the episodes of pain. Numerous experiments have shown that a variety of animals do appear to anticipate painful events. Rats will learn to jump on sight of a warning light which indicates that an electric shock will follow shortly. Monkeys will stay awake (and devlop stomach ulcers) in order to press levers on command to avoid painful shocks. Anyone who has sat in a vet's waiting rooms knows that dogs exhibit clear signs of acute anticipation of impending 'jabs'.

To some extent the evidence that animals do anticipate frightening events seems also to be evidence that they have at least a hazy idea of a future self to which the frightening events will happen. However, it seems to me that self-consciousness is likely to develop gradually out of repeated experiences of fearing, wanting, and so on, the most undeveloped of which need involve no more than, for example, an inexplicable wave of terror or happiness. I don't think that fear necessarily involves any clear idea of what one is afraid of. Even humans sometimes experience mental states such as sudden panic or

[6] P. Singer, *Practical Ethics* (Cambridge: Cambridge University Press, 1979), 78–90, 93–105.

depression which have no relation to ideas of frightening or sad future events, although, in our case, these are normally pathological. If certain methods of animal husbandry make acquisition of learned introspective skills more or less probable, this may raise further moral problems. If, for example, raising animals such as veal calves in social isolation in crates means that they are denied the chance to develop 'higher levels' of consciousness of which they are naturally capable, presumably, if we are sincere in valuing this sort of consciousness, and if destruction of consciousness is an evil, this represents a severe wrong.

The intellectual capacity for self-observation is not the only attribute which ought to be considered if we have to make decisions about the wrongness of killing. For example, it is unlikely that geese understand the significance of death, but killing geese who are members of family groups does cause suffering to the survivors, who search for their lost partners with evident distress. Thus the separation which is intrinsically involved in death may be an evil for many social animals who have little ability to reflect about experience. An animal who does not understand the significance of death may well be able to understand and fear separation from the group. A preference against separation can reasonably be regarded as a preference for life, and fear of separation may be equated with fear of death. Thus a preference utilitarian should recognize that death may be a harm to the individual social animal who is killed even if she has no concept of death or preference for life as such. It is true that, once the animal has been killed, we are not left with a lingering disembodied fear of separation, but the same is true of a self-conscious fear of annihilation of personality. This may be important if we need to make decisions about euthanasia of sick animals. For example, Anne Rasa describes her decision to return an untreatably ill mongoose to die in his own family group:

One of the lowest-ranking of the subadult males, which was just over five years old, had a severe kidney infection, which, despite treatment, went into a chronic stage and it was obvious that the animal would soon die. . . . I decided to return it to its family instead of letting it die alone. . . . The group came out and slept with it on the open floor . . . Again I saw the conscientious licking of the invalid's body, especially by his mother, although at his age, he could not have awakened any maternal feelings in her.[7]

Anne Rasa's knowledge of the way in which dwarf mongooses protect and care for sick group members was clearly importantly relevant to the rightness of what she decided to do.

It would appear likely that killing conscious animals becomes a more serious wrong as the level of self-consciousness involved increases, but that death is a harm even for animals who only possess simple consciousness.

[7] *Mongoose Watch* (London: John Murray, 1984), 262–3.

7

CONFLICTING INTERESTS

The two values . . . —concern for others and desire for knowledge—sometimes come into conflict when the use of the experimental method to obtain knowledge about humans is contemplated. In such cases, the priority of values in our culture is quite clear: No experiment may be done that might harm a member of our species. The appropriateness of this priority seems obvious, since the mistreatment of any person is a potential threat to other people and, hence, to the culture. It is generally accepted, however, that any type of potentially informative experiment may be performed on a nonhuman animal.

Marcuse and Pear, in *Psychopathology in Animals*.

This chapter looks at the facts about the conflict of interests between humans and other animals, and aims to provide an objective examination of the prospect for reforming the way we treat animals whilst avoiding serious harm to human interests.

The central reason why the question of fair treatment of animals is difficult lies in the extent of the conflict between the interests of humans and animals. Humans make use of animals in a wide range of ways, some of which may involve suffering or restrictions on the extent to which those animals can live enjoyable lives. Animals are commonly used in education; entertainment; as sources of raw materials for clothing; to provide motive power; as substitutes for humans in scientific research; to produce foods such as eggs and milk products; and as items of food themselves. Humans may also be in competition with animals for scarce resources, such as food and living space.

I do not aim to provide comprehensive coverage of all the possible areas of conflict between the interests of humans and animals, but merely to outline the major areas of concern. Use of animals in science will be examined in some detail, however, because it is in this area that biological knowledge is most important in deciding what can be done to reduce animal suffering.

Food Production

As suggested earlier there are good reasons for thinking that conscious animals have an interest in remaining alive even if they lack the intelligence to form a distinct idea of death. Hence, if respect for rights depends fundamentally upon respect for the interests of individuals, killing animals would seem to represent a violation of respect for rights. It will be argued in the next chapter that it is too simplistic to say that killing is *always* wrong, and that there may be special factors which render it excusable or justifiable (for example, the euthanasia of an incurably ill and suffering animal, or cases where there is no other way to protect a larger number of individuals). However, the arguments for the wrongness of killing do indicate that there should be a presumption that killing animals is wrong unless there are compelling reasons to the contrary.

According to the figures collected by the FAWC (Farm Animal Welfare Council, an independent advisory body set up in July 1979 by the Minister of Agriculture, Fisheries, and Food and the Secretaries of State for Scotland and Wales under the Chairmanship of Professor R. J. Harrison, Emeritus Professor of Anatomy, University of Cambridge), the total number of red-meat animals slaughtered in Great Britain during the year ending March 1983 was 31,676,000 (3,162,000 adult cattle, 94,000 calves, 14,362,000 pigs, and 14,058,000 sheep). (Some of these were killed for export.)[1] The FAWC estimates that around 450 million poultry are killed annually in Great Britain.[2] In addition an unknown number of minority species, such as goats, rabbits, and horses are slaughtered annually.

[1] *Report on the Welfare of Livestock (Red Meat Animals) at the Time of Slaughter*, Ref. Bk. 248 (London: HMSO, 1984), 4–5.
[2] *Report on the Welfare of Poultry at the Time of Slaughter* (London: Ministry of Agriculture, Fisheries, and Food Publications, 1982), 6.

Clearly, in terms of sheer numbers, this represents one of the most significant areas of concern for those who are interested in the rights of animals. The wrongs involved in killing may be compounded if the methods used cause distress or suffering, or if the conditions in which the animals are reared, transported, and sold do so. In the two reports on livestock welfare at the time of slaughter the FAWC lists numerous points which gave reason for concern that the present treatment of animals at the point of slaughter causes suffering. For example, chickens are normally suspended by their legs on a mechanical shackle line which then drags them through an electrified water bath to stun them, before they are moved on to mechanical throat cutters and finally plunged into a scald tank. The period of suspension is inevitably painful and frightening for the birds, and in addition the Council noted that the stunning procedure was not always 100 per cent effective and that some birds miss the mechanical cutters and may enter the scald tank alive and conscious. Again, in the case of cattle, sheep, and pigs the Council reported that stunning equipment did not always render animals instantaneously unconscious, and that they were sometimes subjected to rough and inconsiderate treatment in the period before slaughter:

We have been concerned at the excessive use of electric goads in some slaughterhouses, both in the unloading process and in the driving of animals, particularly pigs, into the stunning area. In many cases the use of an electric goad was counter-productive, creating confusion and stress for the animals. It was clear that in the hands of some slaughterhouse staff, use of the electric goad became an automatic act in the process of handling of all animals, regardless of whether or not the animals were refusing to move forward.[3]

The RSPCA's report on the welfare of livestock at markets in the UK[4] made fifty-seven recommendations. They estimated that around 20 per cent of all cattle sold in markets are animals who have been rejected from dairy herds because their useful economic life is over, and that many of these animals were suffering from conditions such as chronic mastitis, arthritis, or overgrown hooves and ought to have been transported direct

---

[3] Farm Animal Welfare Council, *Report on the Welfare of Livestock*, p. 12.
[4] *Welfare of Livestock at Markets* (Horsham: RSPCA, 1983).

from farm to slaughterhouse rather than being subjected to the extra stress involved in market sale (p. 21). Like the FAWC, the RSPCA noted excessive use of sticks and goads (pp. 44–5), and they were particularly concerned about the practice of hawking young calves from market to market in order to get the best price (pp. 25–8). Lack of concern for animal welfare at markets is not simply a matter of individual cruelty or indifference: until 1991, when the subsidy was ended, the Ministry of Agriculture required that animals presented for payments be marked by punching a hole in one ear, 1.25 centimetres in diameter and surrounded by six tine indentations, to prevent fraudulent representation of animals (pp. 41–2). Such institutionalized disregard for the sufferings of animals is not likely to help to foster sympathetic attitudes among the market staff and others involved in the transport and sale of livestock.

The use of products which do not necessarily require the killing of animals—wool, milk, eggs, and so on—does not appear to involve such an irreconcilable conflict of interest. At first it might appear that there is no conflict because the products which humans want seem to be obtained without detriment to the animals. However, in practice, humans nearly always inflict incidental harms in the process. Current practices are very far from ideal from the animals' point of view. The majority of egg-producing hens are confined in tiny cages throughout their lives and are killed when their productivity begins to decline, although they could live for very much longer. Laying hens kept in battery cages normally have less than 1 square foot of floor space per bird. They commonly show stereotyped behaviour and suffer foot damage and feather loss.[5] Most male chicks are killed soon after hatching, sometimes by potentially painful methods.[6] Similarly, male dairy calves are usually killed before maturity, often after being kept in severely restricted conditions. See, for example, Marthe Kiley-Worthington's comparison of the behaviour of

[5] See e.g. J. R. Bareham, 'A Comparison of the Behaviour and Production of Laying Hens in Experimental and Conventional Battery Cages', *Applied Animal Ethology*, 2 (1976), 291–303.

[6] W. Jaksch, 'Humane Destruction of Male Chicks', *International Journal for the Study of Animal Problems*, 2(4) (1981) 203–13; Anon., 'Egg Producers Issue Guidelines for Destroying Baby Chicks', *International Journal for the Study of Animal Problems*, (1983), 14–15.

field-raised and confined Friesian calves (mainly bull calves from dairy herds).[7] Kiley-Worthington made her observations in a commercial veal unit run according to the recommendations of the Ministry of Agriculture and within the limits of the Ministry's Welfare Codes of Practice. She notes that older calves were confined in crates measuring 1 by 2 metres, in dim light (dark enough for the observers to require extra light to make notes), permanently tethered by the neck, without bedding, and that they were unable to turn round, lick, or scratch their rumps. They tended to perform more stereotyped actions, such as chewing their pens, head-tossing, rubbing, and scratching than field-reared calves. Severe falls were common when the animals tried to groom their backs and rumps. (Kiley-Worthington suggests that their restricted ability to groom meant that they were permanently itchy.) When finally driven away for slaughter all the calves had some difficulty in walking.[8] Ram lambs are usually killed shortly before maturity (although they are usually reared with their mothers in field conditions), because it is more profitable to keep ewes who can produce wool and a lamb each year.

Reforms which would avoid such severe confinement and stress for animals could be adopted.[9] On a non-commercial basis it is possible to reduce costs to the animals still further, to a point at which there is a genuine balance of costs and benefits from their association with humans. I know several people who keep poultry as pets and who produce enough eggs for their own needs without killing the male birds, but this could scarcely be done on a commercial scale. Possibly there is a need for a complete rethink of our assumption that food production using animals must always mainly be the province

[7] 'The Behavior of Confined Calves Raised for Veal: Are These Animals Distressed?', *International Journal for the Study of Animal Problems*, 4(3) (1983), 198–213.

[8] On 28 Dec. 1986, the Junior Minister for Agriculture announced that legislation banning crate-rearing of veal calves would be introduced during 1987. At this time it was estimated that the 8 veal producers in the UK still using crates were handling 2,000 calves a year (*Farmers Weekly*, 5 Dec. 1986, p. 37). However, this does not affect the conditions in which imported veal will have been produced.

[9] See e.g. Universities Federation for Animal Welfare, *Alternatives to Intensive Husbandry Systems* (Potters Bar: UFAW, 1981); and Paul Carnell, *Alternatives to Factory Farming* (London: Earth Resources, 1983).

of large-scale commercial ventures, and that food-producing animals are not normally companion animals too. Humans can survive and live healthy lives without food produced from animals, although it is probably easier to adopt a lacto-vegetarian diet (eggs, dairy produce, but no meat or fish) than a vegan one (no animal produce except honey). Most humans are probably almost totally vegetarian because they are simply unable to afford meat; a very few live in regions where it is not possible to grow cereal crops so that the only available food source is animals capable of processing grass or marine plankton; the majority of meat-eating humans live in areas where plant food is cheap and readily available.[10] I do not propose to try to tell Eskimos how they ought to live: however, in any country where this book is likely to be read, the balance of interests is overwhelmingly upon the side of food animals and against humans who want to kill them in order to eat them.

Animal Pests

A 'pest' can loosely be defined as any non-human organism which has deleterious effects upon the interests of humans. A. P. Meehan of Rentokil Limited quotes estimates of the losses due to rodents world-wide, ranging from 5 to 20 per cent of total food production, perhaps around 33 million tons per year,[11] although these estimates are probably very rough approximations. In addition, rodents may cause damage by gnawing, and may carry disease, although Meehan considers that rodents are responsible for a relatively small proportion of human infections. Methods of reducing rodent numbers have included poisons, traps, 'sticky-boards' (pieces of hardboard coated with glue in which rats and mice become entangled), chemosterilants, and various kinds of repellents. In the UK the types of poisons available for use are regulated to some extent

[10] Anyone who doubts that a vegetarian diet is cheaper than an omnivorous one in this country can satisfy herself that this is so by by comparing the price of the cheapest complete-protein vegetarian meal (beans on toast) with the cheapest comparable carnivorous meal (pilchards on toast) at any supermarket.

[11] *Rats and Mice: Their Biology and Control* (East Grinstead: Rentokil Limited, 1984), 112.

by the Animals (Cruel Poisons) Act 1963, which requires that they should not cause overt symptoms (mainly convulsions).[12] However, this does not rule out the use of poisons which cause pain but no striking symptoms, such as, perhaps, the anticoagulants (Warfarin, Difenacoum, Brodiafacoum, etc.), which kill over a period of several days by small internal haemorrhages.[13] To some extent, it is possible to control rodent infestation of buildings by proofing so that the animals cannot enter and by avoiding leaving edible waste lying about.[14] It is also possible to reduce rodent numbers by using chemosterilants which prevent breeding, in places where small numbers of rodents can be tolerated, such as sewers and agricultural premises.[15] One humane poison, alphachloralose, is available, which essentially acts like an overdose of anaesthetic.[16]

It is evident that the interests of humans would be severely threatened if rodent control simply ceased altogether, but this does not mean that the animals' interests can be given no weight in rational decision-making on the part of humans. Meehan notes, 'In economic terms, it is preferable to bait with a rodenticidal preparation that kills rats or mice, rather than spend similar amounts of time and money baiting with a chemosterilant' (p. 279). The interests of the animals could tip the balance in favour of a non-lethal method. Similarly, there are probably significant differences in the degree of suffering caused by different poisons and consideration of animal interests could motivate us to choose the most humane types.

It is already possible to control some species by non-lethal methods. In the parish of St Matthews in Cambridge where I live, Joan Court has organized a neighbourhood group of cat lovers, the St Matthews Cat Collective. In addition to acting as a clearing-house for lost and found cats, the collective has organized the neutering of feral cats in the area, thereby preventing colonies from expanding to a point at which they

---

[12] Except in the case of moles, who may be poisoned with strychnine (a painful convulsant) (Universities Federation for Animal Welfare, *Humane Control of Land Mammals and Birds* (Potters Bar: UFAW, 1985), 113–17).

[13] A. P. Meehan, 'Humane Control of Rodents', in D. Britt (ed.), *Humane Control of Land Mammals and Birds* (Potters Bar: UFAW, 1985), 28–36.

[14] Meehan, *Rats and Mice*, pp. 288–92.

[15] Ibid., pp. 277–9.

[16] Ibid., pp. 215–18.

become a nuisance and liable to attract hostile attention from site owners.

Neutering is an effective method of controlling feral-cat numbers[17] and may be cheaper in the long term than killing, but it necessarily involves a greater expenditure in the short term if paid labour is used for the whole process. Local authorities or site owners may consider this a waste of money. As reported by Hayne, 'this is a question of resources as far as the council are concerned and they are certainly not going to put money into the trapping and neutering of cats'. 'If . . . a colony is found to constitute a nuisance or a danger, then trapping and killing is our preferred solution.'[18] However, we feel we have demonstrated that it is possible for relatively small groups of interested people to make themselves responsible for humane control by neutering in their own locality. A pioneering scheme in Islington provides the borough with a neutering service for feral cats at a cost to ratepayers of approximately £10 per adult cat in grant aid to SNIP (Society for Neutering Islington's Pussies), although the total cost is somewhat higher since the authority's paid animal warden provides some additional assistance to the volunteer trappers. SNIP raises additional income by membership subscriptions, and assorted fund-raising activities.[19]

There have been some expressions of concern at the possibility that over-success of neutering could either eliminate feral cats altogether, thus further diminishing the diversity of animal life within urban areas,[20] or else that selective capture of the most docile animals could produce a population of ferocious, untrappable creatures. We considered these possibilities, but felt that, given the small size of the individual colonies, leaving one unspayed female per colony would allow an unacceptably rapid rate of colony growth. Since large numbers of domesticated cats are killed each year in Britain because

[17] Jenny Remfry, 'Humane Control of Feral Cats', in D. Britt (ed.), *Humane Control of Land Mammals and Birds* (Potters Bar: UFAW, 1985), 41–9.

[18] G. E. Hayne, 'Pest Control in Urban Areas: An Assessment of the Problem', in Britt (ed.), *Humane Control of Land Mammals and Birds*, pp. 27, 25.

[19] Information supplied by me by Councillor Aitchison.

[20] Roger Tabor, *The Wildlife of the Domestic Cat* (London: Arrow Books, 1983), 188–92.

they are unwanted, we felt that it would be irresponsible for us to attempt to contain this growth by homing 'spare' kittens. In fact our prime motivation for starting a neutering programme was just this problem of competition for homes between feral and domestic kittens. Before 1982, Miss Ruse, the main cat feeder, organized the neutering of those cats who were easily handled, and was successfully preventing explosive growth of the colonies by homing kittens as soon as they began coming for food. Her efforts in this respect probably explain why we did not have the huge colonies reported by other people.[21]

Attempts to control cats by killing them cause intense suffering to cat feeders, who may be elderly people with no other source of company and too often appear themselves to be treated as pests whose feelings deserve no consideration. Where reduction in cat numbers is *essential* at least a few cats should be neutered and returned—it is not overstating the case to say that the sudden loss of all the members of some elderly lady's feline 'family' could well put her own continued survival in question. I know of no attempts to discover what happens to cat feeders after cats have been removed by trapping, poisoning, or shooting. It is noteworthy that when the whole system of cats plus feeders is considered, painless killing of feral cats clearly poses severe moral problems even on a purist utilitarian hedonistic basis, since cat feeders can experience dread and anguish at the thought of being deprived of their animal companions and also grief and despair after the event, even if cats lack sufficient self-consciousness to worry about the danger.

Analysing the situation in terms of needs poses some problems. A hard-hearted council pest control department might argue that it may be true that cat feeders need *some* cats (and that the council should give their needs as much weight as those of any other citizen), but that they do not need as many cats as there are on some problem sites. I think it is possible to answer this kind of claim by considering the nature of human personal relationships. Most couples probably have a natural need to have children, but before they have had any they do not

[21] Peter Neville, 'Humane Control of an Urban Cat Colony', *International Pest Control*, (1983), 25(5) 144–52; Remfry, 'Humane Control of Feral Cats', pp. 41–9; D. Aitchison, personal communication.

need any particular number. Two children would satisfy the basic need as well as three or four. However, once the children have been born, the parents have a basic need that all of them should survive and be healthy, and their lives will be wrecked if even one dies even if they are still left with a large number of children. A similar kind of account is probably correct for other multiple relationships, such as friendships and relationships between siblings. Similarly, cat feeders have a basic need that their cats are not 'thinned out' by killing. Hence, such actions violate their claims of need.

Whether neutering itself conflicts with the cats' interests to such a serious extent that it represents a violation of their rights is an interesting question. If the choice is simply between neutering the current population and killing them, it would perhaps seem obvious at first sight that neutering is in the cats' best interest. However, someone who believes the cats have important rights to autonomy might claim that they should simply be left alone for their populations to reach a natural balance. If they pose a significant problem for humans (for example, for reasons of hygiene), I do not think that it is possible to say that the humans can be obliged to do nothing to try to resolve the difficulty.[22] Reducing the population by enforcing absolute human non-interference and preventing feeding of the cats would be severely inhumane to both cats and feeders. As argued elsewhere in this book, there is a sense in which animals who are involved in this kind of close relationship with humans can be said actually to form a part of human society, making such enforced abandonment a positive act of cruelty rather than a morally desirable 'letting be'. Considered in these social terms, neutering instead of killing certainly involves less suffering for the cat feeders and also for surviving cats who would be distressed by loss of familiar companions.

Animals in Research —> 151

In considering the question of scientific use of animals it is necessary to be clear that questions about the rights of animals

[22] This idea will be developed more fully in the following chapter.

used are to be kept quite separate from concerns about the role of science itself in shaping modern society. The latter are important and legitimate objects of concern, but, in relation to the question of animal rights, it is the extent to which scientific use causes pain, distress, or death for animals which is central. Thus a person who believes that animals do possess rights is not in any way committed to a general opposition to scientific enquiry, and may fully accept that adoption of objective scientific method has led to important and intellectually stimulating discoveries. For a theory of animal rights in relation to scientific use it is necessary only to consider to what extent these rights ought to limit the kinds of experiment which are ethically justifiable; as indeed scientists have almost universally accepted in the special case of experiments involving the human animal.

It is necessary to point out this distinction because pressure for animal rights has become strongly associated with a more general disillusionment, or even hostility, towards science itself. This is probably in part due to a desire to embrace 'alternative' methods of medical treatment which do not involve the use of animals, and in part to a growing feeling amongst non-scientists that an increasing role of technology and science in our society is inimical to their well-being. It is perhaps important for scientists to try to bear in mind that most people have no real basis for deciding what is the truth about the benefits and harms derived from modern science, since they are faced with opposing sets of statements and little or no evidence. Hostility towards science has scarcely been discouraged by the attitudes of some scientists[23] who have been inclined to adopt ethical systems in which knowledge is seen as the supreme value. Ethics of this kind must of necessity be hostile to the interests of the majority of humans who cannot or do not choose to play a direct role in the increase of human knowledge, since they are relegated to the status of mere servants of those who do.

Science and technology clearly have made life longer, easier, and more pleasant for most people in terms of purely physical comfort and it might seem that the present dissatisfaction with

[23] See e.g. Jacques Monod, *Chance and Necessity* (London: Collins, 1972) 162–7.

what they have to offer is somewhat ungrateful. However, it is also evident that they have contributed to an increasing burden of unease (for which philosophers must also accept some share of responsibility). A world in which the majority of people have less and less chance of controlling or understanding the factors which determine their fate and from which the comforting certainties of the past have been removed must inevitably breed distrust and discontent. For many people the increase in scientific knowledge has not meant greater understanding, but rather the replacement of an untrue but comforting and comprehensible model of the world by terrifying incomprehensibility. This has relevance to the question of animal rights to the extent that justification of animal experiments does seem to me to depend crucially upon the extent to which they succeed in decreasing net pain and suffering: if we would really be better off in blissful ignorance then experiments are not justified. However, we cannot realistically expect to unlearn the knowledge we already have, nor to regain the lost certainties of the Middle Ages. Furthermore, we need to be very careful that an ingrained suspicion of scientific activities, many of which do not involve any harms or injuries to any conscious creature, does not lead us to put an unjustified emphasis upon animal experiments as a source of cruelty to animals. *Eating* animals, which is a much more 'natural' and immediately comprehensible activity, is responsible for a vastly greater quantity of death and suffering than experimentation.

There seem to be some parallels between the problems involved in deciding when our treatment of animals involves unacceptable violation of their rights and some unresolved difficulties about what human society can reasonably demand of its members. Societies generally teach that it is right for men to be compelled to risk their lives in war to defend their country, state, or city, even if they would much prefer to avoid such danger. (Some do have 'conscience clauses' allowing pacifists to opt out—why exactly it is correct to exempt a genuine pacifist, while compelling a genuine coward to participate, is not entirely clear.) On the other hand, suggestions that humans might be compelled to take even slight risks in medical experiments would be greeted with almost universal horror and condemnation. In 1986 British MPs voted against experimentation on human embryos, rejecting it by a substantial majority even though the proposed age limit for experiments would have ruled out all possibility that the embryos might suffer pain. (The proposed limit was 14 days gestation, at which time the embryo has not yet developed a nervous system.) Conversely, abortion is permitted in Britain at relatively late developmental stages when it is just possible that suffering does sometimes occur.

Why should this be so? Where humans are involved there seems to be a virtually universal feeling that research involves peculiar moral horrors. This perhaps should make the animal rights movement's obsession with experiments seem more understandable to scientists. Sir William Paton, for example asks:

So much of it [pain] seems useless, even when one has allowed for the need of warning of bodily damage, or for the necessary price that freedom of action in a dangerous world calls for. It is all the more strange, therefore, that the suffering which carries the hope of reducing future suffering seems to be so bitterly attacked.[24]

It seems likely that the disproportionate concern with experimentation on living animals simply reflects the attitude of society at large, rather than indicating any particular peculiarity of the animal rights movement. Perhaps this imbalance, and the horror at Nazi and Japanese experiments on prisoners, is partly caused because we find it easier to understand killing someone because we hate them (even if we hate for stupid reasons), than killing or injuring for the sake of knowledge, however potentially useful. Partly, also, it stems from a feeling that, if someone is compelled to take part in war, then the responsibility lies ultimately with the enemy. Compelling people to act as subjects of dangerous medical experiments would necessarily mean that responsibility for any harm they might suffer would fall on the researchers, doctors, and legislators of their own community.

It seems to me that experiments on animals could usefully be assessed against a rule concerning what we should consider justifiable if we had to deal with a new disease for which there was no animal model. Would we then content ourselves with

[24] *Man and Mouse* (Oxford: Oxford University Press, 1984), 77.

letting nature take its course? Would we conscript?—Ask for volunteers? How destructive would the new disease have to be before we would consider these methods? How much risk to experimental subjects would we consider was justified in these circumstances?

Clearly, such a test would rule out a substantial fraction of present-day experiments. On the other hand, I suspect that some might be justified by it, and some might be justified with modifications. For example, chimpanzees are used in the production and testing of vaccines against one strain of hepatitis virus. The particular strain involved causes a serious disease in humans, but only a mild one in the apes, so at first sight it seems that these experiments are justified. However, further investigation of what actually happens in hepatitis research reveals that the cost to the chimps is actually very high.[25] Many chimpanzees used for research are captured from the wild. This generally involves shooting mother chimps in order to capture their infants. Thereafter, there is a high rate of mortality during transport,[26] and once they arrive in research laboratories the young chimpanzees are generally kept in isolation in tiny cages. Chimpanzees who become hepatitis carriers are usually killed because of the risk of infecting humans. The comparison test outlined above serves to indicate how we might modify research so that it does not involve such gross costs to chimpanzees. It is possible, though expensive, to keep chimpanzees in large social groups, in which the breeding rate is high.[27] Female chimpanzees commonly leave their natal group around puberty, so that it would not involve unnatural suffering if young females were removed from breeding groups to test groups at this time. Chimpanzees who are hepatitis carriers are not ill, not do they pose a serious health risk to other chimps, though they are a grave danger for humans. Hence it would be possible for chimpanzees who become carriers to be retired to live in social groups. If such a programme were implemented it would seem not to involve

[25] Andrew Rowan, *Of Mice, Models, and Men: A Critical Evaluation of Animal Research* (Albany: State University of New York Press, 1984), 121.
[26] J. Cherfas, 'Chimps in the Laboratory: An Endangered Species', *New Scientist*, 27 Mar. 1986, p. 38.
[27] Frans de Waal, *Chimpanzee Politics* (London: Unwin paperbacks, 1982).

violating rights. One medical research foundation in the USA has actually set up a retirement fund for its chimpanzee research subjects.[28] Like pension funds for human workers, part of the foundation's yearly budget is invested through a trust fund for the duration of the research period so that it should support the chimpanzees to the end of their expected average life span of 45 years. The authors of the paper state that 'Research Institutions have the ethical obligation to reward these animals with a normal life following experimentation by maintaining them for the rest of their lives in indoor–outdoor facilities such as those available at SFBR, in outdoor corrals, or on islands.' Something similar has been attempted by the New York Blood Center's laboratory in Liberia, which has a rehabilitation plan to return its chimpanzee research subjects to the wild after they have been used in the study of hepatitis B.[29] The progress of this project was studied by Alison Hannah. At the time of her paper fifty-seven chimpanzees had been released on to forested islands where there were no indigenous populations of apes. At present they are still provided with supplementary food, and it seems likely that the islands do not provide sufficient wild food to support fully the groups released. It is hoped later to move some of the groups into nature reserves on the mainland of Africa, but this is likely to pose difficulties, since wild groups of chimpanzees normally fight off intruders into their territory. Release into areas where there are no wild apes is problematic because the lack of indigenous populations suggests that there is some reason why the habitat is not suitable. However, Alison Hannah strongly urges that release into semi-wild life on islands is so much preferable to the other possible alternative futures for laboratory chimpanzees that we ought to be prepared to finance long-term supplementary feeding if it turns out that complete rehabilitation into the wild is impossible. In December 1987, the Jane Goodall Institute and the Humane Society of the United States organized an international workshop on improving conditions

[28] Jorg W. Eichberg, and John T. Speck, Jr., 'Establishment of a Chimpanzee Retirement Fund: Maintenance after Experimentation', *Journal of Medical Primatology*, 17 (1988), 71–6.

[29] Cherfas, 'Chimps in the Laboratory', p. 39. Alison Hannah, 'Observations on a Group of Captive Chimpanzees Released into a Natural Environment', *Primate Eye*, 29 (June 1986), 16–20.

for chimpanzees used in research. The recommendations drawn up included the requirement that

Chimpanzees who are no longer useful to biomedical research must be retired and, when necessary, rehabilitated, so that they may live out their lives in conditions that guarantee their psychological well-being, until they die of natural causes.[30]

There is no current research on chimpanzees in the UK, although British researchers are involved in work overseas, and the situation may change if research into AIDS is expanded because chimpanzees are the only animals other than humans in whom the virus can survive, although they do not go on to develop the full-blown disease.[31] It was recently reported that officials from the US National Institutes of Health met congressional aides in February 1988 in an attempt to convince them that thousands of chimpanzees would be needed for AIDS research, and that regulations on use of endangered species should be set aside to make them available.[32] However, the situation is essentially similar for other wild animals (mainly primates) who are used for research purposes.[33]

It seems to me also that if we are really convinced of the importance of medical research we should recognize *some* duty on the part of humans to undertake to serve as subjects in those experiments which do not involve lasting harm. Such humans would need to give up only a minor part of their time and comfort, while research animals used in comparable experiments are allowed little meaningful life and are normally killed once their usefulness is over. Possibly members of the animal rights movement should be prepared to volunteer to replace animals in particular experiments (rather as Kenneth Mellanby's pacifist volunteers undertook to serve as experimental subjects rather than fight as soldiers).[34] This would spare animals and allow the human subjects direct experience of scientific work, giving them a more realistic ability to

---

[30] 'Recommendations to USDA on Improving Conditions of Psychological Well-Being for Captive Chimpanzees', *Alternatives to Laboratory Animals*, 15 (Mar. 1988), 255–9.

[31] Cherfas, 'Chimps in the Laboratory', p. 41.

[32] Editorial, *Alternatives to Laboratory Animals*, 15 (Mar. 1988), 177.

[33] Cherfas, 'Chimps in the Laboratory', p. 38.

[34] *Human Guinea-Pigs* (London: Merlin Press, 1973).

understand that, while there is much wrong with scientific use of animals, random cruelty or sadism is rarely the problem.

It is important to remember that animals are not always used as research subjects simply because it is legal to treat them worse than humans. Where the point of research is to learn something by comparing humans and animals it may often be quite possible to ensure that animal subjects are treated as well as human volunteers. For example, one way of investigating the human genetic system would be to look at all the primate species as a 'natural experiment' in producing variants on a basic primate genome and seeing how they develop. A researcher who was interested in comparing the genes of a variety of primates (which can be done using samples of placental tissue and without injury to the animals) would do this because of the particular value of the knowledge it would yield, not because it would be illegal to study humans by this method.

Some experiments are clearly more justifiable than others: the pharmaceutical trade journal *Scrip* in October 1983 reported that Bayer's new drug Glucobay (acarbose) 'awaits the outcome of long-term animal studies (mid 1984) . . . clinical trials *on humans* have been completed, but the original animal toxicity studies were statistically inappropriate' (my italics). The drug's sales potential was estimated at around 600 million Deutschmarks.[35] This sort of incompetence is inexcusable. One can only wonder how many other experiments, including ones done in this country, are repeated simply because no one could be bothered to take competent advice on the design of the experiment. Clearly it would be desirable to have a legal requirement that all experiments should obtain advice from a statistician (and follow it) so that the numbers of animals used in experiments are reduced to an absolute minimum.

Some apparently innocuous experiments can become unacceptable where the researchers involved lack a proper respect for and understanding of their subjects. Bonnie Beaver quotes a case known to her in which a female cat was allowed to starve to death during an experiment to test cat food flavourings for palatability. This cat found the particular flavourings so

---

[35] *Scrip*, 840 (24 Oct. 1983), 6.

distasteful that she developed severe anorexia and eventually
died.[36] This suffering, as Professor Beaver points out, was
entirely unnecessary since the cat should simply have been
given a different diet when she stopped eating.

Many individuals and societies have concerned themselves
with the question of alternatives to the use of animals in
research in recent years.[37] Some of these organizations
have concentrated on funding 'one-off' projects suggested by
individual researchers, others have instituted more ambitous
long-term investigations; for example, a search for one or more
tests to replace the Draize (eye instillation) test (Hadwen
Trust), and for validation of cell culture toxicological tests
(FRAME).[38]

However, surprisingly there seems to have been little effort
directed towards a broad-ranging view of the aims and
potential of alternatives since the seminal work of Russell and
Burch[39] in the 1950s. This is not the place for such a treatment,
but I hope to indicate lines on which one might proceed, and at
least to provide a short exposition of the reasons why Russell
and Burch's 'three-point plan' is still important.

The 'three Rs' (Replacement of animals; Reduction in the
numbers used per experiment; Refinement of the procedures
so that pain and suffering are minimized) are not sharply
defined categories. For example, a new drug is currently passed
through many different animal tests in the long process of
selection, development, and final safety testing. It is quite
possible that a method which allowed the replacement of

[36] *Veterinary Aspects of Feline Behavior* (St Louis, Mo.: Mosby, 1980), 159.

[37] Among British-based charities alone there are at least 4 organizations
exclusively devoted to funding non-animal research, namely: the Dr Hadwen
Trust for Humane Research, FRAME (the Fund for the Replacement of
Animals in Medical Experiments), the Humane Research Trust, and the Lord
Dowding Fund.

[38] I. F. H. Purchase, D. G. Farrar, and Irene A. Whitaker, 'Toxicology Profiles
on Substances Used in the FRAME Cytotoxicology Research Project',
Alternatives to Laboratory Animals, 14 (1987), 184–242. For more extensive
descriptions of this, and other work on alternatives, see also Dallas Pratt,
Alternatives to Pain in Experiments on Animals, 2nd edn. (New York: Argus
Archives, 1980); Michael Balls, Rosemary Riddell, and Alastair N. Worden
(eds.), *Animals and Alternatives in Toxicity Testing* (London: Academic Press,
1983).

[39] W. M. S. Russell and R. L. Burch, *The Principles of Humane Experimental
Technique* (London: Methuen, 1959).

animal tests in some of the early stages would also mean that final safety testing could be refined by ensuring that animals were very rarely exposed to substances which exhibited unexpected toxicity. By reducing the unpredictability of the results it could also reduce the total number of animals used, for example, in dose-ranging experiments, to ascertain the active dose levels of the drug. If the animals ran no more risk than present-day human volunteers in pre-clinical trials, then the extent of harm to animals' interests would clearly have been decreased, if not eliminated.

This possibility represents a middle way between the view that present-day methods of safety testing must continue because there will always remain a possibility that interaction of effects on different organs and systems in an intact organism will cause toxicity which is not apparent in cultures of single tissues,[40] and those who argue that, if our technology can put a man on the moon, we should surely be capable of modelling a mouse.

Attempts to find alternatives to use of animals in research have largely been aimed either at finding ways to do the research without animal tissue at all (for example, attempts to predict biological activity of chemicals from their molecular structure) or at making use of isolated tissues or organs (for example, predicting toxicity of chemicals from their effects on cell cultures). However, there is a third possible method which needs to be considered if we are rationally concerned to reduce harm to conscious animals, rather than simply applying the verbal rule 'avoid using living animals in experiments': that of using animals which are unlikely to possess conscious experiences.

For example: nematode worms are extensively used in basic research into biological control systems. These worms share many fundamental systems (for example, regulation of development) with all other animals, but have a very rudimentary nervous system comprising only around 200 neurones—in contrast annelid worms such as the common earthworm possess around 10,000 and insects have around 100,000

---

[40] Tam Dalyell, *A Science Policy for Britain* (London: Longman, 1983), 101–3.

neurones.[41] We cannot prove that nematodes have no sensations, but it is evident that we can feel very much more confident that experiments in which they are used do not involve pain than is the case (for example) when an experimental vertebrate exhibits evident pain and fear. In the former case it is stretching belief to claim that pain does occur, in the second only some very dubious special pleading for the uniqueness of the human nervous system can persuade us that the experimenter is not involved in the deliberate infliction of suffering.

The way in which replacement–refinement might work is illustrated by recent successes in the search for alternative methods of testing for eye-damaging potential.[42] It is possible to test potential irritants by applying them to isolated tissue and measuring the amount of damage produced. Tissues used have included bovine eye corneas obtained from slaughterhouses and small sections of gut taken from rabbits killed for other experiments, and the results have proved sufficiently sensitive to discriminate between the irritant potential of adult dandruff shampoo and baby shampoo. Because of the remaining question of possible unexpected ill-effects it is unlikely that these methods will be accepted as the sole method of testing for completely new substances, but clearly if they are used to eliminate all evidently irritant substances before tests on conscious rabbits both the number of animals suffering and the degree of that suffering will be substantially decreased.

Lay people may sometimes falsely believe that animal research is pointless simply because they are not able to understand it, but it should not be forgotten that public lack of understanding of biological science can cut both ways. It is relatively easy for someone to understand the rationale of 'trying it on the dog first', but much harder for lay people to understand how basic research, for example, looking at the

---

[41] David B. Dusenbery, 'Behavior of Free-Living Nematodes', in Bert M. Zuckerman (ed.), *Nematodes as Biological Models*, i (New York: Academic Press, 1980), 128.

[42] C. K. Muir, 'Further Investigations on the Ileum Model as a Possible Alternative to *in vivo* Eye Irritancy Testing', *Alternatives to Laboratory Animals*, 11(3) (1984), 129–34; C. K. Muir, 'The Effects of Surfactant and Hypotonicity on Bovine Cornea *in vitro*: Comparison between Opacity and Thickness', *Alternatives to Laboratory Animals*, 12(3) (1985), 137–44.

diffraction patterns of chemicals extracted from a plant virus, the development patterns of sea urchins, the pigmentation of snapdragons, or the genetics of maize could have medical importance. None the less, these are all studies which won the Nobel prize for Physiology and Medicine, and were of fundamental importance for the development of modern biotechnology. Because of this there will be a tendency for non-scientists who are involved in the control of research funds to favour the more obviously relevant higher-animal work at the expense of projects which use organisms less closely related to humans but which share the characteristic which is of interest for the particular study in question.

The American National Research Council's Committee on Models for Biomedical Research recently published an investigation of the comparative importance of mammalian and non-mammalian systems for basic research.[43] The Committee note that Nobel prize-winners in the biological and medical fields seem to have used non-mammalian systems rather frequently in comparison with the average pattern seen in projects funded by the National Institutes of Health, and they suggest that funding for projects using non-mammals is low relative to the past importance of such projects in contributing to biomedical knowledge.[44] In consequence they suggest that shifting the balance of funding in favour of non-mammalian projects might actually benefit research. Unfortunately their study is mainly concerned with the simple distinction mammal/non-mammal. As was illustrated in Chapter 2, taxonomic categories have to be considered in a more discriminating fashion if they are to be of use in our attempts to minimize suffering. Some invertebrates may be more sensitive than some vertebrates, and some birds may rival even the most intelligent of the non-human mammals. However, the investigation's main importance is probably the way in which it places the use of 'alternatives' to animal research squarely within the mainstream of biology. For example, Maurice Wilkins, Crick and Watson, or Barbara McClintock would probably have been surprised to be told that they were using 'alternatives', but in fact their Nobel prize-

[43] Committee on Models for Biomedical Research, *Models for Biomedical Research* (Washington, DC: National Academic Press, 1985).
[44] Ibid., pp. 44–50.

winning discoveries were all made using non-animal systems as models of the genetic systems of higher organisms (tobacco mosaic virus, cardboard molecular models, and maize plants).

There seem to be two major avenues for replacement of animals in research which need to be distinguished. Firstly, there is the more obvious project of developing direct replacements for specific kinds of animal testing: as, for example, the attempts by the Dr Hadwen Trust to develop ways of testing substances for irritancy using isolated tissues instead of the eyes of living rabbits. Secondly, there may be several quite different approaches to investigating a particular medical problem, only some of which require the use of animals. For example, one research project might aim to improve the treatment of cancers by comparing the effectiveness of different patterns of radiation therapy, using animals with induced tumours as models of human patients. Another project might look at the basic biology of tumour cells in culture, in the hope of discovering more about what makes them different from normal cells, which in turn could lead to new ideas for treatments which kill cancer cells but leave normal ones unharmed. (The contrast between these two kinds of experiment can perhaps be compared to the difference between the work of Edison and Einstein, the former progressing largely by highly methodical trial and error,[45] the latter by asking questions about the basic nature of the world.)[46] The second experimental plan is not really an alternative way of doing the first one, but can better be described as a different strategy aiming at the same general goal. Of course some basic research does use living animals, and much applied medical research does not. The facts documented by the report suggests that a shift towards funding of basic research using simpler organisms and models would be beneficial to medical science, but they by no means prove that there would be no costs involved at all if use of whole animals were to be abandoned immediately.

Considering the whole spectrum of animal use in science and technology it seems to me that it is helpful to group purely technological use at one end of the range and pure science at the other, with various intermediate categories. This subdivision

---

[45] Ronald W. Clark, *Edison* (London: Macdonald and Jane's, 1977).
[46] Jeremy Bernstein, *Einstein* (Glasgow: Collins, 1973).

directs our attention to the different strategies for reform which will be effective in the different areas.

1. Technological use of animals for standardizing known substances; for example, measuring the strength of insulin preparations by injecting samples from a batch into mice and following the effect on their blood sugar levels.[47] Testing for the presence or absence of known contaminants; for example, administering samples of injectables to rabbits so that contamination by gram-negative bacteria will be revealed by the fever reaction which their toxins provoke. Diagnosis of disease; for example, injecting mice with test material to detect the presence of live leprosy bacteria.[48] Production of substances required for experiments; for example, growing monoclonal antibodies in the ascites fluid of mice implanted with cancerous cells.

2. Use of animals for routine toxicity testing; for example, administering different amounts of the substance to groups of mice to determine the quantity which will cause 50 per cent mortality (LD50 test). There will usually be standard tests which the researcher will be required to do in order to produce data which will satisfy the various regulatory bodies responsible for allowing new products on to the market, but there is some scope for individual ingenuity, for example, in devising screening tests to allow early rejection of substances which are too toxic to be used; or in improving experimental methods and design, for example, reducing the numbers of animals needed for reproducible results by making conditions more uniform so that there is less random variation to obscure the effects of the test substance—for example, if some animals are suffering from a sub-clinical infection they may be less able to survive dosing.

3. Development of new medical treatments or improvement of existing methods; for example, screening substances for effectiveness in treating depression by administering them to rats previously dosed with reserpine (which causes subjective depression in humans and stereotyped gnawing behaviour in

[47] 'HPLC in Place of Animal Assays', *Alternatives to Laboratory Animals*, 14 (1987), 129.

[48] 'Screening for Drug Sensitivity', *Alternatives to Laboratory Animals*, 15 (1988), 182.

rodents, which is abolished by effective anti-depressants). There is considerable scope for individual creativity, athough this is still an area where it is possible to 'do science' by following standard protocols.

4. Basic research into biological function: at the 'lowest' end Kuhn's puzzle-solving behaviour and at the 'highest' his 'revolutionary science'. At this end of the scale there is a high degree of individual autonomy, although some workers at the lower end will still be engaged in rather standardized pursuits which vary only in detail from one individual to another (for example, gene sequencing is now becoming a technological routine).

5. In addition there is a fifth category, education and training, which may involve any of the other four (trainees may be, for example, technicians, new researchers, medical workers, and so on), but typically will involve the demonstration of known facts and close supervision and control of students.

These distinctions are important to those of us who want to see an end to research which causes suffering because they define areas of science and technology which need different strategies of action.

Animal use in category 1 is repeated, standardized, and directed to simple, well-defined ends. (Is the solution contaminated? Is the insulin of normal strength? Is the infection present?) Individual research workers have little influence or choice about the methods used, which are normally standard tests demanded by government regulators. If an alternative method which does not use live animals can be devised to the satisfaction of the regulatory bodies it can then rapidly replace the live-animal method. Because the aims are simple and well-defined there are good prospects for developing replacements. (For example, it is now possible to test for gram-negative contamination by mixing the test solution with a drop of blood from the king-crab (*Limulus*). If the mixture coagulates then contamination is present. One king-crab can provide enough blood for several tests and can be returned to the sea unharmed. Physical tests of the strength of insulin preparations are approaching the degree of sensitivity at which they can replace the rabbit test.)[49] A tissue culture method has been officially adopted by the Animal Disease Research Institute of Agriculture in Canada to replace the mouse inoculation test for rabies diagnosis. It is estimated that this will save an average annual use of 30,000 mice.[50] Coenraad Hendriksen of the Laboratory for Control of Bacterial Vaccines in The Netherlands has recently pointed out that biological standardization is a most important area for research into alternatives to use of animals because not only are the numbers of animals involved very large but the distress inflicted upon them is often severe.[51] When the numbers of experimental animals used in The Netherlands in 1986 were categorized by distress rating (slight, moderate, or severe), it was found that 23.7 per cent of all animals used were believed to have suffered severe distress, but that 37.3 per cent of the animals used for production, testing, and standardization of biological products fell into this category. Category 1 procedures accounted for 20.8 per cent of all animals used in 1986. At the Laboratory for the Control of Bacterial Vaccines, use of a cell culture method as an alternative to the guinea-pig lethal-potency test for standardization of diphtheria vaccine reduced the number of animals used each year by about 3,000. Previously groups of guinea-pigs were inoculated with different dilutions of the vaccine and later 'challenged' with diphtheria toxin to measure the protective effect. The effectiveness of the vaccine was measured by the number of guinea-pigs who survived at the different dilutions. The tissue culture alternative still involved some animal use: mice were immunized with the vaccine, then killed and the amount of toxin-neutralizing antibody in their blood measured by its ability to protect sensitive cells from damage by diphtheria toxin. However, fewer animals are needed because more information can be obtained from each individual, and none has to suffer a painful death from the effects of diphtheria toxin. Category 1 use is repeated again and again until an alternative is found, so any developments in this area will save

[49] 'HPLC in Place of Animal Assays', *Alternatives to Laboratory Animals*, 14 (1987), 129.

[50] *Alternatives to Laboratory Animals*, 14 (1986), 61.

[51] Coenraad F. M. Hendriksen, 'Laboratory Animals in Bacterial Vaccine Control and Some Opportunities for Replacement, Reduction and Refinement', *Alternatives to Laboratory Animals*, 16 (1988), 125–36.

large numbers of animals over the years. Replacement is generally the only possible course of action for category 1 use since the substances in question are of immediate importance and the tests cannot simply be abandoned.

In category 2 the question posed, 'How toxic is this substance?', is more indefinite. A substance may become toxic only after processing by the liver; it may affect some organs and not others. It is possible that a substance which seems quite harmless when cultured with any individual body tissue may behave differently in the whole body. However, it is also the case that substances which are evidently toxic to individual cells, or are chemically similar to other toxic substances, are likely to be toxic to whole animals (and to humans). It is possible that the general public may never be willing to give up the psychological reassurance of 'trying it on the dog first', but improved screening tests which reject highly toxic substances before animals are used would be a significant step in reducing suffering. Again, individual researchers have relatively limited influence on the basic nature of the experiments[52]—there may even be standard tests laid down in detail by the British Standards Institution, such as BS 5736, 'Method of Testing for Skin Irritation by Solid Medical Devices', which specifies length of application of the test material, number and species (rabbits) to be used, and the way in which the results are to be graded on a scale of 0–4 for redness and swelling of the skin.[53]

The distinct areas of activity are also significant if we are attempting to achieve political change. The British government has frequently responded to requests for state-sponsored research into alternatives by stating that it is considered that alternatives are best developed in the course of normal research by urging scientists to seek non-animal methods whenever possible. However, as I have shown above, this is not possible for scientists and technologists working in categories 1 and 2, who have very limited control over their activities and may well lack the training and facilities which would be needed to develop new non-animal methods. Students and

[52] D. R. Laurence, A. E. M. McLean, and M. Weatherall, 'Conclusions', in Laurence, McLean, and Weatherall (eds.), *Safety Testing of New Drugs* (London: Academic Press, 1984), 163–8.
[53] *Alternatives to Laboratory Animals*, 11 (1983), 1–2.

trainees (category 5) can scarcely be expected to develop alternative methods although they can and do make their feelings known about unnecessary demonstration experiments. It is evident that a research programme targeted at category 1 would have a good prospect of success in eliminating animal use, while targeting of category 2 could realistically hope to achieve significant reduction in the numbers and severity of experiments on live, conscious animals. Furthermore, because experiments in these groups are largely standardized according to regulatory requirements, it is necessary for the technologists and scientists to obtain approval for the substitute methods; the final decision rests with government.

In fact, it may be the case that public demand for safety testing has actually caused unnecessary proliferation of animal experiments. Laurence *et al.* note:

One way of assessing the value of toxicity testing to contemporary standards is by historical comparison. . . . Publications [concerning drugs introduced prior to 1960] describing the pharmacological properties or first clinical use of such drugs sometimes contain a brief account of toxicological studies, involving for instance dosing a few animals of two or three species for as long as a month . . . Penicillin was evidently used immediately for life saving purposes, because the quantity originally available was minute. . . . It is interesting to note that much testing was done on cell cultures *in vitro* . . . One may wonder if the present regulatory demands would crumble if a situation were to recur in which such urgency of therapeutic need was unmistakable.[54]

Mary Tucker *et al.*, in the same volume, point out that the results of tests on the anti-cancer agent Tamoxifen were initially confusing because this compound behaves like a weakly active version of oestrogen.[55] At relatively low doses it acts against tumours which require oestrogen for continued growth by competing with natural oestrogen produced by the body; however, at very high doses it may actually promote tumour growth since massive quantities of weak hormone activity have the same effect as smaller amounts of high activity. Thus traditional toxicity studies which give animals

[54] *Safety Testing of New Drugs*, p. 166.
[55] M. J. Tucker, H. K. Adam, and J. S. Patterson, 'Tamoxifen', in *Safety Testing of New Drugs*, p. 158.

massive doses of test substance in the hope that this will simulate the effects of long-term human use are quite inappropriate for this kind of drug.

Research in category 3 involves more individual responsibility, but it is still the case that the majority of workers will be proceeding using standard techniques. Charitable or government-funded research into alternatives could reasonably be expected to develop some new non-animal methods (for example, it has proved possible to screen compounds for possible pharmaceutical activity by monitoring them for effects on plant enzyme action). Research into the way in which known drugs act by binding to molecules in the body can make it possible to design better ones which fit the required binding site more exactly.

In category 4 it is probably true that scope for providing researchers with ready-made alternatives is limited. However, even revolutionary science does not operate in a vacuum. An individual researcher's methods will be influenced by many factors including her previous scientific education, the availability of alternative model systems, and the attitude of society to use of conscious animals.

It is possible to identify areas where further development of techniques and equipment would open up broader potential for non-animal research. For example, MRI and MRS (magnetic resonance imaging and spectroscopy) make it possible to produce cross-sectional and three-dimensional images of internal organs and to investigate their chemical constituents without harm or discomfort to the patient.[56] Currently the usefulness of these methods is limited by the speed and memory size of existing computers, which cannot handle data fast enough to cope with (for example) the image of a beating human heart, without some loss of definition. Removal of these limitations is a soluble technical problem which would open new areas of basic research into human biology. Because MRI and MRS can be done without harmful effects they also have the potential to enable research into animal biology which is acceptable from the point of view of the interests of the animals involved.

Developing equipment of this kind is a research engineering project in its own right, not something which is likely to be

---

[56] Gary P. Zientara and Leo J. Neuringer, 'Medical Imaging for the 21st Century', *Perspectives in Computing*, 8 (1988), 14–24.

done in the course of a biological experiment. However, once a facility is available it will be used by researchers who might otherwise have had recourse to invasive experiments on animals. Because such equipment has a generalized usefulness (unlike, for example, the very specialized value of the *Limulus* test for contamination), it will serve to replace the use of animal experiments in basic research without in any way limiting scientific freedom of thought.

Quantitative analysis of the historical development of science can enable more precise targeting of effort in the search for alternatives. It has been demonstrated that technological developments tend to follow a logistic curve, that is, improvements are initially exponential but there then follows a period of slowing down as a particular method approaches its limits of usefulness, and finally another period of exponential growth when a new breakthrough provides further scope for refinement.[57] It is of little value to spend money and effort on a process which is approaching its physical limits, but targeting the period of early development when large improvement is possible and the new ventures which may open up fresh prospects for exponential growth would enable resources to be spent most effectively.

Some specific areas of research which have promise for reducing use of animals can be distinguished.

Organ and Tissue Culture

It is possible to keep animal cells alive outside the body if they are provided with a culture medium containing appropriate nutrients. Such cells may then be used for experimental studies of functions which do not depend upon the interaction of organized tissues of the whole body. Normal body cells will eventually stop dividing, age, and die, and need to be replaced at regular intervals, either from human sources, such as biopsies or fresh placental tissue, or by using tissues from animals killed for that purpose or in the course of other experiments. However, some cultured cells become 'transformed' by a process which appears to be similar to the change which occurs

[57] Derek J. de Solla Price, *Little Science, Big Science* (New York and London: Columbia University Press, 1963), 20–32.

when body cells become cancerous. These cells will grow and divide indefinitely in culture and can be used to model the reactions of normal cells, since many of their systems remain unaltered by the transformation.

Cell cultures can play a useful part in pre-screening of substances for evident toxicity, thus reducing the number of experiments in which living animals will be subjected to damaging substances. If a substance will damage cells which are exposed to it, it is reasonable to suspect that it is likely to damage body tissues in the whole animal and the substance can therefore be rejected without testing on live animals. FRAME has organized a large-scale investigation of the toxicity of various chemicals to lines of transformed cells, in an attempt to validate their use in screening and to discover the most effective ways of measuring cell damage.[58]

Some whole tissues and organs can be kept alive for some time after removal from the body, and these can be used to investigate their function in isolation. Such studies normally involve the killing of animals, except where unwanted human tissues such as placentas, organs donated for medical research, or ones removed in the course of medical operations are available. However, it is often possible to use tissues from animals killed for other purposes, for examples, eyes obtained from slaughterhouses, so that these studies do not require more animals to die than would have done so anyway. Because isolated organs are more accessible to monitoring equipment than those remaining in the whole body it may be possible to reduce the total number of animals used by making a large number of measurements on each organ instead of a few measurements on many individual animals. Use of isolated liver in this way has now reached a commercial stage, in which compounds such as pharmaceuticals, pesticides, industrial chemicals, and herbicides are pre-screened to discover how they are metabolized. Since the liver is the main organ in which metabolism of foreign chemicals occurs, this goes some way towards overcoming the objection that only whole-animal

---

[58] R. J. Riddell, D. S. Panacer, S. M. Wilde, R. H. Clothier, and M. Balls, 'The Importance of Exposure Period and Cell Type in *in vitro* Cytotoxicity Tests', *Alternatives to Laboratory Animals*, 14 (1986) 86–92.

studies can predict whether chemicals will be metabolized to more poisonous products in the body.[59]

The potential of organ culture for reducing use of animals is illustrated by a system for evaluating chemicals for teratogenicity (tendency to produce deformities of the foetus) developed by Oliver Flint of ICI. In this system, cells from early rat embryos are exposed to the test chemical in culture and then monitored to see if their ability to differentiate (develop into the form characteristic of their type) is affected. For example, if a teratogenic substance causes malformations because it prevents nerve-cells from sending out their processes it will normally also inhibit process formation in cultured nerve-cells. Because only a few cells are needed for each test culture one embryo can provide the cells for several tests, and, although animals are killed, they need not be caused to suffer. Only three animals are required for a complete screen of a new substance, compared with 300 for a conventional study in which pregnant animals are dosed with the chemical. The US Food and Drug Administration is currently investigating the test to see whether it should become part of the required screening programme for new pharmaceuticals.[60]

Mathematical Models

Extravagant claims for and against the potential of mathematical models have been made, particularly since the advent of relatively cheap, high-speed computing. On the one hand, it is evident that any mathematical model requires some input of factual data in order to generate predictions which can be of any use in the real world, while on the other it is not entirely fair to state that such models can do no more than simulate known facts. As early as 1917 D'Arcy Thompson was demonstrating that mathematical modelling of the morphology of animals and plants could provide entirely new evidence about the ways in which this morphology is likely to be

[59] Anon., 'News and Views', *Alternatives to Laboratory Animals*, 14 (1986), 62.

[60] Anon., 'Award for *in vitro* Teratogenicity Test', *Alternatives to Laboratory Animals*, 14 (1986), 2.

determined.[61] Population geneticists regularly use mathematical models and computer simulations to discover the effects of varying selection pressures, genetic drift effects, and breeding systems on evolutionary trends.[62]

Recently, computer simulations of various systems of the body have been developed to replace animal experiments and demonstrations in teaching. This kind of simulation is possible because existing knowledge about the systems can be expressed in terms of mathematical equations, for example, ones which relate the concentration of dissolved carbon dioxide in the blood to rate of breathing. If these equations are specified as exactly as possible then the student can perform experiments by typing in changes in some of the variable quantities, for example, an increased figure for the quantity of oxygen in the inspired air, and seeing how these changes affect the behaviour of the whole simulation. Where basic knowledge about how the living system works is already extensive it may even be possible to use such programs in research, for example, by simulating how a variety of treatment regimes might affect patients and selecting the most promising ones for clinical trials.[63] Such models can provide a useful way of instructing students in the concepts of physiological systems and experimental design, since learning is much more effective if students are actively interacting with a model rather than merely memorizing facts. However, they cannot, of course, provide experience of the handling of fresh biological material, and may still need to be supplemented by work with actual tissues. The models certainly offer prospects for reducing the numbers of animals used in biomedical education, perhaps to levels at which it is possible to confine use to ones who have been killed for other purposes or died naturally. It is already common for veterinary students to get most of their experience in dissection using animals which have been sent for post-mortem examination.

---

[61] *On Growth and Form*, revised edn. (Cambridge: Cambridge University Press, 1942).

[62] There is an entire journal, *The Journal of Theoretical Population Biology*, devoted to mathematical modelling of the behaviour of populations.

[63] See e.g. C. J. Dickinson, D. Ingram, and K. Ahmed, 'The *Mac* Family of Physiological Models', *Alternatives to Laboratory Animals*, 13 (1985), 107–16.

Quantum Pharmacology (Quantitative Structure-Activity Relationships or QSAR)

The chemical elements are built up in a regular way by serial addition of more fundamental particles—protons and neutrons to the atomic nucleus; electrons to the surrounding orbitals. Their chemical properties depend upon the way in which these particles are arranged, and most significantly upon the distribution of electron density surrounding the component nuclei. Because of this, it is possible to predict the chemical behaviour of a particular kind of atom or molecule by solving the Schrödinger wave equation which describes the form of the electron cloud which surrounds the nucleus. Nuclear charge is much less significant in determining chemical activity because it is the outer electron 'shells' which interact when two atoms or molecules approach one another.

In practice, complete solution of the wave equation is possible only for very simple systems such as the atoms of hydrogen and helium. Larger, more complex molecules of the kinds involved in biological reactions can be calculated only approximately. However, for the smallest organic molecules (such as ammonia, NH_3) some chemical properties can now be computed to a greater degree of accuracy than is possible by experimental measurement. Larger molecules require a higher degree of approximation if they are not to take up a huge amount of expensive computer time, and hence the results obtained are less reliable. At present theoreticians are trying to solve the added problem posed by the fact that biological reactions do not occur in chemical isolation, but within complicated solutions in water or lipid, which modify the behaviour of the electron clouds. Meanwhile, it is possible to control for such effects to some extent by performing calculations on molecules which extend existing chemical series[64] and, rather than trying to predict their behaviour from scratch, noting whether the calculations suggest they will be more or less reactive than series members whose behaviour is

---

[64] Compounds which have similar molecular structure and which are varied in a more or less regular way, e.g. by addition of extra $-CH_2$ groups to a chain of carbon atoms ($-CH_2-CH_2-CH_3$ vs. $-CH_2-CH_3$), are known as members of chemical series.

known. Any errors due to the approximations used will be the same for all members of the series, so should not significantly alter their relative behaviour.

Chemical activity depends not only upon the way in which a molecule's component atoms are linked together, but also on its *conformation*, the way in which a chain of atoms folds or rotates to produce a variety of possible shapes. Atoms which are brought together by such folding or twisting can interact through their electron orbitals, by repulsion or attraction, causing some conformations to be more stable (less energetic) than others. This behaviour, like interaction with other molecules, can be calculated from the wave equation, and the frequency of the different conformations predicted. This is particularly important in biological systems where reactions, especially those involving enzymes, often involve changes of molecular conformation as important stages. Computer graphics are frequently used to make 'maps' of the potential-energy contours of the possible conformations of molecules so that the least energetic (most stable) can be determined, and to construct 3D pictures of molecular shapes. The latter can be used to obtain a visual impression of possible molecule/recept or behaviour in the body. Theoretical calculation of the structure and activity of biological receptors is still in its infancy, since few have been isolated and characterized. Some theoretical information may be obtained by studying the structure of compounds which are known to act on a particular kind of receptor. Since there must be a relationship between receptor structure and chemistry and the properties of the molecules which act on it, either to block its normal biological action or to stimulate it, this kind of study can illuminate possible receptor structures. Some important biological receptors, like the dopamine receptor of nerve-cells, have many known activators (agonists) and blockers (antagonists) which may be compared to discover their common properties, and hence the molecular properties which are essential for pharmacological activity. Sometimes rather similar receptors have importantly different biological function. One example is the different types of dopamine receptors, whose malfunctioning is probably important in schizophrenia (over-activity) and in Parkinson's disease (under-activity). In the control of schizophrenia it is important to ensure as far as possible that blocking of dopamine receptors is limited to the ones which are over-active, otherwise patients may experience involuntary muscle movements rather similar to Parkinsonian tremors. Comparison of currently available anti-schizophrenic drugs can help theoreticians to design compounds with maximum anti-schizophrenic activity, and minimum production of tremor i.e. ones which are designed to 'fit' the rceptor involved in schizophrenia as specifically as possible.

Calculations have now reached a stage where it is possible to use them to design compounds which are likely to have useful biological activity, although they do not by any means yet approach a state of perfection in which they give absolute prediction of chemical behaviour. However, they are already making it possible to move away from the old system of 'molecular-roulette' pharmacology in which new compounds were synthesized virtually at random and then 'screened' by dosing animals and looking for activity of any sort. Since pre-screening for activity 'consumes' more animals than safety testing of drugs, quantum pharmacology offers significant promise as an alternative to experiments on living animals. By helping to select chemicals which are unlikely to have significant negative effects it is also possible that suffering during the period of safety testing can be reduced.[65]

Biotechnology

The technique of genetic engineering for the production of proteins of medical or economic importance has considerable potential for reducing dependence upon the products of slaughtered animals. Genes for a particular protein are inserted into bacterial DNA, generally with bacterial control genes attached so that the cells can be 'switched on' to protein production. The bacteria are then grown in culture solution and the proteins they produce are harvested. An idea of the potential of this technology can be gained from the fact that the company Genetech was able to produce as much of the hormone somatostatin from 2 gallons of bacterial culture as could be made by processing half a *million* sheep brains

---

[65] W. G. Richards, *Quantum Pharmacology* (London: Butterworth, 1983).

(normally obtained as a by-product of the meat industry).[66] Somatostatin is of medical importance in treating children who are unable to produce it themselves and would otherwise suffer from acromegalic gigantism (excessive growth of the bones). The ability to replace the livestock slaughter industry's products is probably of relatively small importance at present while a majority of Westerners eat meat, but is likely to become more so if vegetarianism increases. Ultimately, genetic engineering of bacteria (and probably also yeasts) to produce animal proteins could also eliminate the need for slaughterhouse products for culturing some particularly fastidious types of cells which presently require added animal serum; replace the use of animal-derived hormones, such as insulin, in medicine,[67] and perhaps provide cultured nutrients for obligate carnivores like pet cats. It could help to provide safer vaccines which would require less animal testing.[68]

Film and Video Recordings

Where it is necessary to train students or technicians to recognize the appearance of various conditions, for example, the characteristic effects of nicotine poisoning, but they do not need to practise any kind of manipulation, it is normally now possible to replace the use of demonstrations with living animals by recordings. This means that large numbers of animals will be spared because the demonstration experiment need be done only once.[69]

No method of testing can provide absolute safety. Isolated human tissues may not possess the interactive metabolic pathways which convert a relatively harmless substance into a toxin, and they can give little evidence about rates of absorption and excretion, which will affect the maximum concentration produced within the whole body. Animals may

---

[66] J. Cherfas *Man Made Life* (Oxford: Blackwell, 1982) 152.

[67] Some genetically engineered human insulin is already in use.

[68] For a brief account of the current status of genetic engineering of eukaryote, see R. J. Warr, *Genetic Engineering in Higher Organisms* (The Institute of Biology's Studies in Biology, 162 (London: Edward Arnold, 1984).

[69] Anon., 'Video Demonstrations Replace Animals', *Alternatives to Laboratory Animals* (1986), 62–3.

possess metabolic processes which humans lack, or may absorb or excrete material at vastly different rates. For example, if a compound is rapidly lost via the kidneys an animal may be capable of tolerating relatively greater doses per unit of body-weight than a human with slower excretory processes. It might be expected that results from primates would provide reliable estimates of risk, but this is not always so. The rodenticide pyriminyl was expected to be very safe because it appeared to be much more poisonous to rats and mice than to other animals. The acute LD50[70] for *Rattus norvegicus* (brown rat) is 4–13 milligrams per kilogram body-weight, while 60 mg/kg are needed to kill cats; 500 mg/kg poison dogs; and 2,000–4,000 mg/kg are needed to poison monkeys. In the event, when the rodenticide was released for use there were 'numerous' incidents of accidental human poisoning and it appears that by some quirk of metabolism this supposedly rodent-specific poison is also specific to humans, who can be killed by tiny amounts of the compound.[71] Of course this does not prove that animal tests are no use at all in predicting toxicity; in very general terms it is likely that what poisons one animal will often also be harmful to another species, in this case the human one. The example does illustrate one reason why using large numbers of animals to establish very accurate LD50 values is unjustifiable. If animal toxicity gives only an approximate indication of likely human toxicity, then the animal toxicity measure might as well be an approximate one. This requires fewer animals than accurate determination for statistical reasons. If a group of animals are all given equal quantities of a substance, then individual differences between them will cause their reactions to vary. If a large number of animals is used, then these differences will tend to cancel out, and the average response will be an accurate measure of the susceptibility of that species. If only one or two animals are tested, it may happen by chance that the ones picked are individuals who are much more, or much less,

---

[70] The LD50 (Lethal Dose 50%) is the dose level at which half of a group of animals can be expected to die after treatment. An acute LD50 is the dose which will cause death relatively shortly after treatment; chronic-poisoning tests measure longer-term effects.

[71] Meehan, *Rats and Mice*, pp. 241–2.

susceptible than average. Thus, the results of the experiment can give only a rough estimate of the toxicity of the test chemical for that particular species. Statistical calculations can tell us exactly how great an accuracy is possible for a given size of test group. To minimize the harm inflicted in safety testing, therefore, we should not aim for greater accuracy than will actually be of use in predicting safety. High-precision toxicity values will not be useful if their relation to human toxicity is only an approximate one.

Clearly it also matters that we should try to reduce current suffering of animals who are used in research. The Association of Veterinary Teachers and Research Workers has recently published preliminary guidelines on the assessment of pain, stress, and distress in animals.[72] Of particular interest are their comments on the evaluation of stress, which they consider can be classified on a scale of increasing severity:

1. Physiological stress within the range of normality, in which the animal makes fine adjustments of which she is probably mainly unaware (as when our heart rate rises during mild exercise).

2. Overstress. The animal is required to divert significant resources into maintaining normal function, but is still little aware of the process.

3. Distress. Substantial effort by the animal is needed to adapt to the stressor; she is likely to be aware of, and caused to suffer by, the process.

Clearly, helping researchers to become more aware of the suffering of animals, and giving them more information about preventing and alleviating it is one way in which the distress suffered by laboratory animals may be reduced. As pointed out by Russell and Burch, the mere fact of reducing the numbers of animals involved in experiments will also help to decrease suffering because it means that more space, time, and care can be devoted to individual animals.

Sheila Silcock of the RSPCA has recently produced a discussion article on ways in which experiments can be refined

---

[72] J. Sandford. R. Ewbank, V. Molony, W. D. Tavernor, and O. Uvarov, 'Guidelines for the Recognition and Assessment of Pain in Animals', *The Veterinary Record*, 22 Mar. 1986, pp. 334–8.

to reduce stress, discomfort, and suffering; for example, pointing out that skin irritation studies could be made much less severe if a transparent dressing were used. This would enable the researchers to end the experiment as soon as any inflammation became apparent instead of waiting a specified length of time before examining the treated area, by which time irritation may have become serious.[73]

However, it is important that efforts to increase the comfort of laboratory animals should be based upon a sincere concern for their welfare rather than the desire to obtain 'a "cleaner" and more standardized research tool'.[74] B. A. Baker describes group pens which allow only 0.825 square metres of floor space (2.5 metres pen height) for adult baboons of around 15 kilograms, and individual cages allowing 0.524 square metres floor space and 1.06 metres height, not much more than a decent-sized rabbit hutch.

Companion Animals

In the case of companion animals, over-population caused by irresponsible breeding is probably the major cause of conflict between human actions and the interests of the animals. (In 1985 Battersea Dogs' Home in London alone killed 8,721 unwanted dogs,[75] and in 1989 the RSPCA estimated that at least 1,000 healthy dogs are killed every day because they are unwanted.)[76] It might be argued that preventing pets from reproducing is itself an abnormal restriction of their behaviour, or even that pet-keeping in general involves so much interference with animals' natural activities that it ought to be discontinued. Investigation of the behaviour of the wild species which are ancestral to domestic pets, and of feral animals which have reverted to the wild, can help us to

[73] Sheila R. Silcock, 'Refinement of Experimental Procedures', *Alternatives to Laboratory Animals*, 14 (1986), 72–84.

[74] B. A. Baker, 'Breeding Baboons in Cambridge: A Retrospective Assessment of the Degree of Success Achieved over Seven Years', in Universities Federation for Animal Welfare, *Standards in Laboratory Animal Management* (Potters Bar: UFAW, 1984), 265–71.

[75] Anon., 'News and Reports', *Veterinary Record*, 119 (7) (6 Aug. 1986), 142.

[76] *RSPCA Today* (Spring 1989), 18.

discover the extent to which such concern is justified. For example, studies on wolves have shown that it is normal for the dominant male and female in a pack to prevent subordinates (who are usually adult offspring or siblings of the dominant couple) from reproducing.[77] Humans take the place of pack leaders and/or parents in their relationships with dogs so it would appear that they are not imposing significantly more severe restrictions than the animals could expect to experience in a free-living state. Of course this kind of evidence could not justify human actions which caused suffering (we would never accept methods of killing food animals which were as cruel as those used by chimpanzees, for example).[78] However, where the question is one of justifiable restriction of animals' liberty it seems reasonable to take account of the nature of restrictions which would be likely to exist in animal societies living independently of humans. Study of the behaviour of free-living animals can also help us to choose the course of action which involves least disruption of natural social patterns. For example, people often avoid having female cats spayed because they feel it is unkind to deprive them of the experience of motherhood, but do not feel that there is anything wrong with removing all the kittens provided that they are found good homes. Cats tend to be perceived as solitary animals who will lose interest in their young once they are weaned. Even disregarding the inevitability of such unrestrained reproduction leading to the killing of animals for which no homes are available, study of cats in a free-living state shows that this view is based on a kind of inverted anthropomorphism which assumes their capacities are less than they actually are. Male offspring typically remain in the colony until they are 1 to 2 years old, when they are driven out by the dominant tom. Females normally remain for the rest of their lives, and co-operate in the rearing of subsequent litters. (For example, related females have been observed to suckle one another's offspring, and females without young may share prey with weaned kittens.)[79]

[77] Erik Zimen, *The Wolf*, trans. E. Mosbacher (London: Souvenir Press, 1981).

[78] J. Goodall, *The Chimpanzees of Gombe* (Cambridge: Mass.: The Belknap Press of Harvard University Press 1986), 290–2.

[79] Michael Allaby and Peter Crawford, *The Curious Cat* (London: Michael Joseph, 1982).

These observations suggest that the aim of providing a more natural social life for pet cats might be better achieved by keeping small groups of related animals, allowing some members to reproduce occasionally, than by allowing single females to have repeated litters which are removed at weaning.

Similarly, in the case of small rodent pets the quality of the animals' lives may be improved by matching the species of animal with the conditions which can be provided. For example, adult hamsters maintain solitary territories, except when breeding, and it is no kindness to attempt to keep them in groups. On the other hand social species like guinea-pigs and rats have an evident desire for company, although it may not be possible to keep adult males in the same enclosure unless they have been reared together. It is unlikely that these small animals are distressed by the idea of losing their freedom as a human would be, so that they will be happy provided that their cage is large enough to allow normal activity. However, cages would need to be much larger than those normally sold in pet shops if they were to achieve the ideal of a size which the animal does not recognize as confinement.

Humans may cause suffering to companion animals by irresponsible selective breeding for attractive appearance. In some ways, this may be a three-way conflict between pet, pet keeper, and show breeder. Companion animals have an interest in a long, healthy, and non-neurotic life, which is shared by their keepers, and tends to be in conflict with the desire of breeders to produce animals with fancy show characteristics, such as the exaggerated gait of modern German shepherd dogs, or the long coat and squashed face of a Persian cat. There is no intrinsic reason why selective breeding of companion animals should not be diverted away from the interests of show breeders in more rational directions.

Other definable areas of animal use include the production of furs, cosmetic substances, work, and entertainment. The former clearly involve extreme costs for the animals, who are likely either to be reared in unsuitable, confined, farm conditions or to be caught by painful methods such as the gin trap (outlawed in Britain, but not in the USA or in Canada). The use of animals to do work or to entertain people will be discussed more fully in Chapter 10; however, it is sufficient here to note that such use may be entirely in keeping with the animal's needs and interests (for example, working sheep-dogs) or may involve unacceptable suffering (for example, keeping wild animals confined for circus performances), and that individual examples need to be examined on their own merits.

Our principle for deciding what is justified and what is not should be that use of animals is only justified if there is good reason to believe either that the animals derive sufficient benefits to compensate them for any restraints or harms imposed by humans, or that using animals is the only way to preserve humans from death or significant harm. In the second case we are obliged to do all we can to reduce costs to the animals to a minimum.

8

BEASTS, SAINTS, AND HEROES

It might be argued that forgoing personal benefits in order to avoid causing injury to non-humans is an act which is good, but not obligatory, and, on this view, animals may be said to be creatures who possess moral status, but not rights.[1] To accord even this much weight to the interests of non-human animals represents an increased regard for their moral standing compared with some current attitudes (see below). However, I believe it is possible to demonstrate that accepting that at least some animals possess at least some definite rights to moral attention involves fewer violations of our intuitive perceptions than an obstinate insistence that only humans can be rights-holders.

I think opponents of the idea that animals can have even correlative rights, founded on our duties towards them, would rarely deny the proposition that a world in which suffering in general was diminished would be preferable to the present state of affairs, i.e. good. What they do dispute is the idea that any individual person could have a duty to make sacrifices in order that this desirable state of affairs should be realized, and the idea that human interests should not necessarily be given overwhelming weight in the calculation of general welfare. I think it is unlikely that anyone would seriously dispute the suggestion that a world in which animal suffering was reduced

[1] As, perhaps, Rawls, who says, 'While I have not maintained that the capacity for a sense of justice is necessary in order to be owed the duties of justice, it does seem that we are not required to give strict justice anyway to creatures lacking this capacity. But it does not follow that there are no requirements at all in regard to them, nor in our relations with the natural order (*A Theory of Justice* (Oxford: Oxford University Press, 1972), 512). Rawls thinks that we have duties of compassion, but not justice, towards animals, and although he does not specifically talk about rights it is arguable that his position is essentially equivalent to that of someone who believes that these are restricted to humans or, even more specifically, to humans who are competent as moral agents.

would be preferable to present conditions if this could be achieved without even the slightest inconvenience to humans.

Some opponents of animal rights may genuinely believe that the consequences of ending animals' exploitation would be an actual increase in *net* suffering. The most extreme version of this would be the belief that all products of animal suffering, including such apparent trivia as furs, cosmetics, and hunting for sport, are actually essential to the happiness and welfare of human beings, whose refined sensibilities would suffer a degree of anguish unknown to mere animals if deprived. Were this view factually correct, I think it would have to give consequentialist supporters of animal rights some concern. However, the existence of humans who willingly choose to experience such deprivation tends to disprove the view that all products of animal exploitation are of overwhelming importance. Thus, in these cases, if our treatment of animals ought to depend upon balancing the competing interests involved, it seems that we must either do without products which involve significant harms to animals, or reform the systems of production to a point where these harms are eliminated.

If we consider more serious human concerns, for example, food production, it is unlikely that the benefits of intensive farming genuinely outweigh the costs in animal suffering, since such farms use more edible vegetable protein and energy than they produce as meat and eggs. There are, perhaps, only two really plausible areas of genuine conflict between basic interests of humans and animals: firstly, competition for resources (pest control), and, secondly, the use of animals in medical research. Here there may be a genuine conflict between harming a few individuals and allowing many to suffer naturally originating evils, or between harming animals and allowing humans to suffer. However, the arguments for permitting harmful use of animals in these circumstances could equally well justify hurting small numbers of human beings in extreme cases where there was no other way of saving greater numbers of people, so they do not necessarily tell us very much about the question of whether animals have *any* rights. It could be the case that both animals and humans have rights, but that it is permissible for these to be overridden in extreme circumstances. Conflict between natural rights is discussed by J. Teichman, who suggests that we should distinguish between cases where a right is properly overridden by another more significant one and those where it is improperly overridden. (In the second case she says that it is best to say that the right continues although it has been violated.)[2] Possibly it might be helpful to think of conflict between rights as situations where a greater right should be said to negate or cancel a lesser one. If both X and Y have moral rights by virtue of their nature as sentient beings and these rights happen to be in conflict it seems reasonable that we should consider the total situation and act in favour of the individual whose right is greater. If we continue with an arithmetic metaphor in which one right cancels out another we may also think it is reasonable to say the individual whose right is the greater incurs a corresponding obligation (for example, to make reparations if this becomes possible) because of the effects of the lesser right, which exists even though it becomes negated when the total situation is considered.

Thus it could be a moral duty for a diabetic to refrain from eating meat, but permissible for him to use insulin although this is generally obtained from slaughtered animals and standardized by tests on mice. However, I think such a person would also have a duty to do his best to obtain the synthetic insulin which is now available in small quantities, and to support attempts to find methods of standardizing batches of the hormone by chemical means.[3]

---

[2] *Pacifism and the Just War: A Study in Applied Philosophy* (Oxford: Blackwell, 1986), 77 ff.

[3] In fact standardization of insulin is a good example of the possibilities for progressive reduction of animal use in experiments. This hormone was originally assayed by testing the quantity required to produce convulsions in mice. This test has now been replaced by one which involves injecting mice with the insulin preparation and measuring the fall in blood sugar level which results. This means a reduction both in the severity of the procedure (there is no need to continue to a level of hormone which causes convulsions), and in the number of mice required (from about 600 to around 130 for each batch of insulin). Methods of standardizing insulin by High Performance Liquid Chromatography (a test of the physical properties of the insulin solution) and by measuring the extent to which the test material will compete with a standard of radioactively tagged insulin for a receptor on cultured human cells are currently under development, but have not yet been fully validated ('Reduction of the Use of Animals in the Development and Control of Biological Products', *Lancet*, 19 Oct. 1985, pp. 901–2).

Species loyalty does count for something, and therefore we should (in extreme circumstances) prefer to sacrifice an animal rather than a human. However, we still have a duty to discover whether the choice really is as stark as this, or whether there is a way to avoid sacrificing either human or animal. If we do not accept this, then it seems to me that we are putting ourselves in essentially the same false position as someone who simply assumes that new cosmetics are essential to women's happiness or that it is essential for us to develop new food colourings, flavourings, and preservatives.

In a situation where we possess factual knowledge which enables us to compare the amount of suffering caused by exploiting animals with the amount prevented, we might well be justified in saying that a person who abstained from exploiting animals at serious cost to himself was going beyond the demands of duty. (The question seems to be complicated by the fact that some humans experience strong emotional bonds with particular animals and are quite frequently prepared to endure hardship in order to protect them. In this situation, safeguarding the animal's welfare seems to be treated as a positive desire, rather than an obligation.) None of these considerations can detract from our obligation not to harm animals for trivial reasons in non-extreme circumstances.

In might be said that a decrease in net suffering does not make a particular course of action obligatory, although it does mean that it is praiseworthy, and we are therefore not positively obliged to avoid harming animals. This seems hard to justify in the case of negative obligations, such as refraining from killing animals to eat if vegetable food is available, since we very definitely do not think that merely refraining from harms towards innocent human individuals is a cause for praise. It is clearly obligatory. Since discussions of animal rights focus mainly upon the justification for inflicting *suffering*, and since this sensation is an evil for both humans and (many) animal species, it seems unlikely that the reason for the difference in attitude lies in a difference in the act of inflicting suffering (which is essentially the same whatever the victim's level of intelligence). If the difference is claimed to lie in the social attitude of the dominating individual, i.e. he can emphathize with other humans, but not animals, then it is not easy to say why this should not also apply to (for example) the relations between different human races. If kindness to dogs is supererogatory, and need be performed only by people who happen to like dogs, then why is it not supererogatory to avoid racism? Would those who believe it is supererogatory to avoid harming animals think that equality for members of other races was supererogatory if it should ever be proved that there were measurable inherited differences in IQ between different races? Indeed why do we not conclude that avoiding causing harm is supererogatory in *all* cases where it is possible for us to cause harm and get away with it? It could perhaps be argued that only human beings can be contracting members of human society, and that members of other species are 'outside the pale' for this reason. However, this leaves the anomaly of those domestic animals who function in many ways as members of human society; and those groups of humans (for example, native South American Indians) who belong to quite separate contracting groups from our own, but clearly do have rights to decent treatment. Furthermore, there are some classes of human beings such as the severely mentally retarded or the deranged who are not capable of forming contractual relations with any other humans. If extreme demands are made with respect to capacity for shared communication before rights shall be accorded, what should be our view of claims (for example) that men and women attach different emotional content to some words which describe feelings? If this turned out to be true, it would be no more possible for human beings of opposite sexes to be *certain* about the significance of one another's emotions than for them to be certain that a crying rat or dog actually feels pain. What they do have good reason to believe, however, is that other beings disclose clear preferences about what happens to them. It is these preferences which demand our moral attention, and we should avoid becoming too much entrapped in speculation about the precise qualities of the emotions of other individuals.

A final argument which might be suggested is that consideration for animals is supererogatory in the sense that only a minority of humans are 'called' to promote this behaviour. As I have argued above, it seems implausible that abstention from harming can be supererogatory except in extreme circumstances, since we definitely do accept that humans have negative rights, although it is controversial whether they can have positive ones. There do not appear to be any relevant differences between sentient animals and humans which could justify the belief that harms to the former are of *no* moral significance and we should at the least be prepared to avoid causing such harms in cases where nothing of greater moral significance is at stake.

However, it could well be the case that active promotion of animal rights (as opposed to simple abstention from direct harms) is not obligatory for the greater part of the human race. Since the world is faced with so many other causes of suffering it seems very reasonable that there should be no compulsion to specialize in relief of animal suffering (although possibly strict utilitarians should consider that it is their duty to do so because of the immense numbers of animals who suffer). If there is no obligation to perform acts which actively promote acceptance of the idea that animals have rights, then those people who do this, or who assist sick, injured, or stray animals, can be said to be performing actions which go beyond what strict duty requires. Ironically, this is probably a more favourable estimation than is usually accorded, since such behaviour is often regarded as not being *beyond* duty but rather as a self-indulgence which prevents concentration on the 'real' obligation to succour human beings.[4] F. L. Marcuse complains:

The height of irony regarding the 'murder' of baby seals happened to one of the writers (F.L.M.) in 1970 when he was living in the United States. He was asked to sign a petition requesting the Canadian Government to cease and desist from this inhumane practice. It so happened that at this very time the United States was using napalm to reduce the Vietnamese population![5]

In fact the assumption that people who care about the way animals are treated are indifferent to human welfare seems not

---

[4] Naomi Lewis, Children's Books Editor for the *Observer* notes: 'one of the most revealing features of human attitudes to animals is the absence of all gratitude; the resentment when any concern is shown. *"Why don't you think about humans!"*' ('Animal Trap', *The Vegan*, NS, 1(2) (1986), 7).

[5] F. L. Marcuse and J. J. Pear, 'Ethics and Animal Experimentation: Personal views', in J. D. Keehn (ed.), *Psychopathology in Animals* (London: Academic Press, 1979) 315.

to be borne out by the facts. Several attitude surveys and opinion polls have shown that people who are involved with animal welfare bodies tend to be more, rather than less, concerned about issues such as aid to the Third World than their uninvolved peer groups.[6] They were also significantly less likely to be racially prejudiced.

Resentment of concern for animals is not new, and a variety is recorded in the Buddhist myth in which a prince complains, on being rescued from drowning by the Buddha, in the form of a hermit, 'This rascally hermit pays no respect to my royal birth but actually gives brute beasts precedence over me' (because the hermit–buddha had rescued three small animals, who were also swept away by the flood, and proceeded to warm them first because they were weaker than the healthy, well-fed prince).[7] This is genuine 'speciesism', the insistent demand that the 'royal birth' of human beings should entail an automatic right to precedence over the needs and interests of non-humans. Too many people still oppose considerate treatment of animals not in order to secure benefits for humans, but simply on the principle that if humans cannot have them then no one shall. It seems that they would prefer to see 'animal lovers' spend their money on themselves rather than give it to the RSPCA, and appear to feel that if suffering is to be relieved at all then it must be human suffering, but that it would be perfectly acceptable to relieve no suffering at all.[8] Thus, recognition of some instances of regard for animals as

[6] Richard H. Thomas, *The Politics of Hunting* (Aldershot: Gower Publishing, 1983), 157–73.

[7] Jataka 73, in *Jataka Tales*, ed. H. T. Francis and E. J. Thomas (Cambridge: Cambridge University Press, 1916).

[8] This attitude will be familiar to anyone who has been involved in collecting money on behalf of animal welfare charities. Such collectors are frequent recipients of critical comments of the form, 'I wish you took some interest in children's welfare.' Presumably those making this criticism would not make similar remarks to people who were merely engaged in private entertainment: e.g. going to the cinema. In fact one major charity, the People's Dispensary for Sick Animals, actually words a good deal of its advertising material so as to maintain the polite fiction that PDSA workers are not directly concerned with the welfare of the animals they treat, but primarily with the needs of poor or elderly pet owners who would suffer distress and hardship if there were no source of free treatment for their animal companions. This, in a sense, merely passes the responsibility one stage back: clearly *some* humans must be directly concerned for the welfare of some individual animals if the situation is to make sense.

supererogatory might mean an actual enhancement of their position on our current moral scale.

The situation becomes more difficult when sparing animals is liable to mean that genuine hardship will result for people *other* than those advocating restraint. It is in many ways unfortunate that the claim that non-human animals have rights has become so firmly associated with absolute opposition to all use of animals in biomedical research. It should be relatively easy to convince an unprejudiced person that animals ought not to be harmed merely to provide things which we find pleasant, and that vegetarianism is therefore a moral necessity. It is far more difficult to argue that animals have such an absolute right not to be harmed that we should be prepared to forgo our own self-preservation and, still more, that we should do so on behalf of other humans. Of course, some nominally 'medical' research does not have life-saving potential and could be rejected on the basis of the kinds of arguments that one would use to support vegetarianism (that it is unjust to subject sentient beings to non-trivial pain where the likely consequences are only trivial benefits to a smaller number of beings, and so on). However, it would be foolish to deny that there are some experiments which do produce non-trivial benefits for humans. If one argues that these too must be abandoned if we accept the concept of animal rights, it seems to me that the resulting moral position is very similar, in several interesting ways, to that of people who adopt an ideal of strict pacifism towards other humans. Proponents of either position must justify themselves against accusations that what they are really saying is that they are prepared to accept any consequences, however bad, rather than get their own hands dirty by compromising their moral positions. Presumably it would be fair enough to choose to do this if one were the only individual involved. If, however, the situation is one in which other people suffer because of my refusal to use violence in their defence, inaction seems much more difficult to justify. I propose to argue that 'extremist' positions can at least be shown to be consistent and coherent. Of course, demonstrating that beliefs are not positively irrational is not the same as providing proof that everyone ought to be prepared to adopt them. However, it seems to me that it is at least possible to argue that ideas which can be shown to be held seriously, with good reasons, and with a high degree of internal consistency, ought not to be simply dismissed out of hand.

There appear to be two possible types of belief in absolute rights. Firstly, it may be thought that these are rights which must never be infringed whatever other consequences may result. ('Let justice be done though the Heavens fall.') Secondly, someone may be prepared to concede that it would be justifiable to infringe rights if this were the only way to avoid more serious consequences, but may consider that, in practice, this will never happen. For example, he might agree that it would have been justifiable to assassinate Hitler if this could have been done without causing other bad consequences, but claim that, in practice, it is not possible to isolate individual acts of violence in this way, and that absolute respect for rights is a principle of rule utilitarianism. There is perhaps a degree of overlap between the two categories. If someone believes that rights must always be respected, and can never be forfeited, because they are based upon moral principles enjoined by God, he may say that this means that humans must obey these moral principles at whatever cost and that they ought not to be swayed by any possible consquences. However, I think such a person would also be likely to say that God approves these principles because they are good, and that the ultimate reason for this is that God (being omniscient) knows that obedience to the principles will lead to his creatures' ultimate happiness. It seems to me that he must accept some variant of this, otherwise he will become involved in a circular argument in which God approves moral principles because they are good and moral principles are good because God approves of them. Even God must have some kind of reasons for approving moral principles, and the subjective feelings of his creatures seem as plausible as any. Even an intuitionist who claims that he 'just does' see that it is always wrong to override rights presumably believes that the result of such actions are in some sense worse than those of inaction, although he may think that the badness lies in the actions themselves, rather than in identifiable consequences. ('The end never justifies the means.')

Cheyney Ryan considers that the pacifist position stems chiefly from an intuitive recognition of the common humanity of self and 'enemy' which makes killing impossible, rather than from any chain of reasoning.[9] It is valuable as a reminder that we should not be too ready to thrust groups of people outside the pale of moral consideration, but dangerous in that it denies the loyalties to family and country which drive people to kill in their defence. In his view, the pacifist's point of view could best be summed up in the question addressed to those who kill, 'How could you bring yourself to do it?' It is possible that we have some innate inhibitions against killing members of our own species, and that fatal violence becomes much easier if we see our opponents as 'not quite human'.[10] If this is true, then individuals may be subject to natural variation in the strength of their inhibitions and some may simply find it impossible to kill once they have made the intuitive leap which is necessary in order to see the hated 'enemy' as a conspecific. Clearly it would make no sense to condemn such people for refusing to kill since they actually *cannot* do it.

If variation in inhibitory strength is less important in the genesis of pacifism than this intuitive step which causes us to recognize that the enemy is also human, other possible consequences may also affect the balance of good and evil. It is possible that forcing people to kill against their natural inhibitions could eventually mean that they might cease to have normal inhibitions towards members of their own group.

People who advocate absolute rights for animals tend, on the whole, to be individuals who also have personal relationships with some non-human animals (personal observation). I do not think it is unreasonable to suppose that similar processes are active in the prodution of their convictions. Once one has seen a *particular* animal as a friendly individual it no longer becomes easy to regard another individual animal as a laboratory 'tool'. In this respect, the advice of the senior physiologist James Pascoe to young biological scientists, urging them to overlay their 'gut reactions' to painful experiments with reason,[11] appears psychologically unwise. There may well be

[9] 'Self-Defence, Pacifism and the Possibility of Killing', *Ethics*, 93 (Apr. 1983), 508–24.

[10] K. Lorenz, *On Aggression* (London: Methuen, 1966), 67–71.

[11] *Attitudes to Experimentation on Living Animals: Science, Ethics, Law: A Personal View* (Leeds: Education Subcommittee of the Physiological Society, 1983).

people who for their own sakes ought not to become involved in painful or fatal experiments on animals. Attempts to force them to continue by accusations of sentimentality are both unwise and unkind.

Whether we like it or not, a large percentage of modern Western people probably have become accustomed to treating a variety of animal species as social partners, through childhood experience of pets.[12] This means that situations which require them to inflict harm upon these creatures may involve overriding or deadening the social impulses of sympathy and inhibitions against hurting which are essential to their relationships within their own species. Human beings are incomparably the most dangerous species of animal which has yet inhabited this planet, and any factor tending to produce individuals with desensitized inhibitions may perhaps pose an incalculable threat to us all.

These sorts of considerations may convince us that there exist some people whose desire to behave non-violently ought to be respected. However, they do not establish that non-violence is tenable as a rational moral position which could theoretically be universalized. People who do claim to be pacifists or to believe in absolute rights generally, I think, want their beliefs to be seen as more than simply a personal quirk.

A consequentialist absolutist would probably argue that we are all really aware that the individual we propose to harm is a sentient being much like ourselves. Modern war also inevitably involves the killing of innocent third parties such as young children and we know it does. However 'unsentimental' we may be about animals, we cannot avoid knowing that they have not done anything to deserve being used in experiments. Under normal circumstances we do believe that it is wrong to harm the innocent, and we believe this on moral grounds, rather than because our feelings are particularly involved. Deciding to override this sort of moral belief in the cause of greater good is liable to weaken our respect for moral

[12] It has been estimated that 50% of households in the EEC include one or more companion animals of some kind (Peter Messent and Steve Horsfield, 'Pet Population and the Pet–Owner Bond', in *The Human–Pet Relationship* (Vienna: Institute for Interdisciplinary Research on the Human–Pet Relationship, 1985), 9–17).

prohibitions on harming in general. For example, painful experiments in Britain were originally allowed (under the 1876 Cruelty to Animals Act) only if they were likely to advance physiological or medical knowledge which might help to save life or reduce suffering. It is arguable that modern justifications of the 'necessity' of developing and testing substances like food colourings using animals do represent a lowering in our moral standards. (In contrast to the Cruelty to Animals Act, which, theoretically, permitted only medically relevant experiments, the Act which replaced it states that the Animal Procedures Committee, which advises the Home Secretary, 'Shall have regard both to the legitimate requirements of science *and industry* and to the protection of animals against avoidable suffering' (my italics).[13] The original law's insistence on the moral importance of balancing suffering caused against potential relief of suffering seems to have been diminished. If this kind of change is a widespread effect than the true net consequences of violence may be less favourable than the short-term ones. This may apply even in the case of morally guilty humans, such as those responsible for initiating violence, since killing or injuring them may have deleterious effects in terms of diminished respect for human life and so on. (To those who argue that animals cannot be innocent in the fully moral sense of a human being who has made a conscious choice not to do evil, I would reply that, in a modern war waged for genuinely justifiable reasons, virtually all adult civilians must accept *some* guilt through complicity in their government's actions. Obviously there are degrees of guilt, but it is unlikely, for example, that the Nazis could have retained power in Germany if a majority of the population had not passively supported Hitler's policies. In this kind of situation it can fairly be argued that the few people who do take a positive stand against the regime are likely to feel that external attacks upon it (short, obviously, of global annihilation) are beneficial to them, rather than the reverse unless they are themselves pacifists. If they die during such attacks they seem to be more like casualties on the right side than innocent victims. However, we do still feel that the killing of children, who are innocent in the same way

[13] Animals (Scientific Procedures) Act 1986 (London: HMSO, 1986), s. 20(2).

that animals are innocent, is a particularly evil part of modern warfare, and we do not think that this evil is mitigated by children's inability to make moral choices not to involve themselves in the waging of war. Indeed, we might go further, and say that children who become caught up in fighting at ages where they cannot fully appreciate the consequences are innocent in spite of their physical actions, precisely because of this lack of full understanding.

Non-materialists would probably also wish to add that violence may be liable to cause more than merely *physical* effects. For example, if the use of violence is somehow spiritually corrupting, then using violence to avoid physically bad consequences could still produce worse net effects even if the physical results were a success. Jan Narveson argues that pacifism is inconsistent because if you are opposed to violence it makes no sense to refuse to employ a small amount of violence to prevent a greater amount.[14] But, if I, by killing a single murderer, prevent two other people from being killed by him, net consequences will not necessarily be optimal if the corrupting effect of willingness to take life is very important and I produce a situation in which two people have become corrupted instead of only one. This argument gains force where the person who must be killed or injured to save other lives is, in fact, innocent, but I do not find it fully convincing. True, we must all die sometime, and it may be fair to say that the moral aspects of what we do are more important than adding or subtracting a few years here or there. It does not seem to me that quality of life can so easily be dismissed. If those who could have been saved have close emotional relationships with other individuals, then this second group of people will have the quality of their lives very severely impaired, and for a longer period than would be expected in the course of nature. (Spouses can normally expect to be bereaved only relatively late in life, children normally survive their parents, and so on.) For someone who is confident of survival after death this may not be an insuperable objection. If the bereaved survivors participated in the moral choice to reject violence, this might

[14] 'Violence and War', in Tom Regan (ed.), *Matters of Life and Death* (New York: Random House, 1980), 109–47.

produce spiritual growth or other important valuable effects. However, it does not seem to me that it is reasonable to expect other people to accept this decreased quality of life if they do not believe that personality survives. Hence it does not seem reasonable for non-materialists to expect materialists to be persuaded by this sort of argument, or by some kind of modification of Peter Singer's argument that we desire to act ethically because this adds meaning to life,[15] if net physical consequences are clearly *very* deleterious. Some kinds of moral nobility may be just too costly (although it should be pointed out that some people who do not accept absolutist positions *vis-à-vis* animals or guilty people do unconditionally reject experiments upon human embryos even when it is quite certain that these are not sufficiently developed to feel any pain). I think it is possible to challenge Narveson's claim that pacifism is logically inconsistent on more materialistic grounds. The whole point of non-violence is to reduce the *total* level of violence and this claim is not refuted by demonstrating that pacifism does not 'work' in individual specific cases. I think it is possible to demonstrate that this is not an impossible idea, by considering the question of the right to carry weapons for self-defence. This applies in the USA, but not in Britain. Clearly there must be individual cases where lives are lost which might have been saved if the victims had been able to defend themselves using firearms. However, it is also very probable that *in general* restrictions upon the number of lethal weapons which are in circulation does help to diminish the total quantity of violence in British society compared to that in the USA. Pacifism might therefore be a strategy which is successful in reducing violence *in toto* (even if there remain numbers of people who will not comply), perhaps by subtly altering attitudes so that there is a generalized increase in the threshold of provocation which is necessary to generate a violent response.

If many of the higher animals are in fact perceived by humans as actual members of human society, if lowly ones, then it might well be the case that increased reluctance to cause harm

---

[15] *Practical Ethics* (Cambridge: Cambridge University Press, 1979), 216–20.

to them could also be a factor in promoting a generalized increase in consideration and reduction in readiness to act violently. Thus there may be no inconsistency in denying that it is right to cause a small number of animals to suffer in attempts to prevent the suffering of larger numbers of individuals through disease. It is perhaps significant that those religions which have the strongest injunctions against violence towards human beings also preach non-violence towards animals.

This kind of down-to-earth reasoning does mean that it is necessary to admit that violence *would* be justified if it could be proved that the net physical consequences were optimal. This is a less drastic position than the one suggested by the non-materialist argument I outlined above and, in some senses, is less fundamentally 'pacifist'. The pacifist who argues *for* non-violence as a method of promoting peace, rather than *against* violence as the ultimate defilement, is making a different kind of claim (although of course he does regard violence as a particularly serious evil). Similarly, attempts to show that respect for rights is justifiable in terms of self-interest, or even in terms of net reduction in suffering, do seem bound ultimately to run into some difficulties. We tend to feel that there will always exist the possibility of circumstances in which breaking our moral code turns out to be the best way to save our own skin. I am not certain that this kind of problem is soluble, even in the case of basic *human* rights, if we are determined not to appeal to additional factors, such as moral nobility.

Narveson additionally argues that pacifism is a form of free-loading. He claims that society is fundamentally an organization of individuals who agree on a common policy of mutual defence against aggression by maverick individuals or external groups. Hence the individual pacifist cannot have a 'right' to refuse to participate without placing himself outside society. It is only fair to say that some pacifists have also recognized this as a moral difficulty. Gandhi records that he came to the conclusion that, while living in the British Empire under the protection of the British Fleet, he was necessarily participating in violence, since that protection depended upon the Fleet's potential violence. He concluded that it would be possible to escape from this participation only by leaving the Empire or by seeking imprisonment through civil disobedience.[16]

Gandhi also noted similar problems in his policy of non-violence towards animals. Pests might be driven away rather than killed, but this merely represented a way of shifting the difficulty to adjacent farms: in effect a form of violence towards his neighbours. Since the pest animals would be killed in their new habitat his attempt at non-violence towards them would not in fact diminish the total quantity of violence towards animals. If he did not kill animals himself, but also did not forbid members of his community to do so, he merely evaded guilt by shifting the responsibility for taking life on to other people. If he ordered all his community to refrain from killing pests this could be a form of violence since they might then suffer the effects of reduced crop yields or increases in the numbers of dangerous animals such as snakes. In fact Gandhi may have been wrong in believing that it is not possible to protect crops effectively by scaring animals back to natural wild habitats where they will not compete with human needs for food. It is interesting to compare the two symposia on the humane control of vertebrate pests published by UFAW in 1969 and 1985. The former is almost entirely concerned with the question of how to ensure that killing is done as painlessly as possible; sixteen years on, almost all speakers devoted at least some space to control by contraception, exclusion from premises, deterrence, or other non-lethal methods.[17] But, in general terms, I think he correctly recognized a serious problem of attempts to live without violence.

Similar problems face the modern campaigner for the rights of animals. Control of vertebrate pests may now be possible by prevention of reproduction, rather than killing, but these new methods have themselves been developed at the cost of some suffering to innocent animals. Even people who boycott all modern medicines benefit to some extent from treatments given to others which curb the incidence of disease (although

[16] M. K. Gandhi, *An Autobiography; or, The Story of My Experiments with Truth*, trans. Mahadev Desai (Ahmedabad-14: Navajivan Publishing House, 1940).

[17] Universities Federation for Animal Welfare, *The Humane Control of Animals Living in the Wild* (Potters Bar: UFAW, 1969); D. Britt (ed.), *Humane Control of Land Mammals and Birds* (Potters Bar: UFAW, 1985).

there is a considerable body of evidence which suggests that public health measures are a more successful means of reducing disease than medical treatment for those already sick).[18] Yet more questions arise when specific cases of conflict between the welfare of animals and humans are considered. The polio vaccine used in most countries is composed of weakened, but not killed, organisms which are mildly infective to contacts of the children who are inoculated. This vaccine has to be tested by inoculating it into monkeys because it is possible for the organisms to revert to the virulent form which causes paralysis. Vaccine made of killed polio virus could be tested in culture using human cells because it is only necessary to check whether the virus is really dead, not to examine whether it is capable of damaging the nervous system of a primate, since dead virus is certainly safe. However, health authorities prefer the live vaccine because it gives immunity even to people who refuse to be inoculated since they are likely to catch the weakened virus from children who have received the vaccine. Many of the people who refuse vaccination do so because of objections to the use of animals in production and testing of the vaccine but ironically this means that they are indirectly reinforcing the decision to use the live vaccine. Respect for human rights prevents the use of coercion to ensure that the whole population is treated with killed vaccine. Animal liberationists who accept vaccination now in the hope that killed vaccine will become acceptable are directly contributing to the present toll of monkey deaths. Further, the vaccine is normally given during childhood. Do parents have the right to refuse to give their children the best available protection from this crippling disease?

Gandhi concluded finally that

Ahimsa (not-harming) is a comprehensive principle. We are helpless mortals caught in the conflagration of himsa (harming). . . . A votary of ahimsa therefore lives true to his faith if the spring of all his actions is compassion, if he shuns to the best of his ability the destruction of the tiniest creature, tries to save it and thus incessantly strives to be free of the deadly coil of himsa. He will be constantly growing in self-restraint and compassion, but he can never become entirely free from outward himsa.

[18] D. Melrose, *Bitter Pills* (Oxford: Oxfam, 1982), 10–14.

Gandhi, I think, would accept Narveson's criticisms of the pacifist as free-loader (although he would perhaps claim that the value of the possible decrease in violence through personal training in avoiding violent behaviour and thoughts justified this), but there are possible replies to Narveson's charge. It is possible that pacifists actually ought to be prepared to opt out, either by civil disobedience or by physically removing themselves from the possible benefits of other people's violent actions. Some modern pacifists do consider that they ought to avoid contributing to war preparations by ensuring that they do not earn a wage which is sufficiently great to pay tax. Secondly, it might well be argued that society does depend for its survival upon more than defence against human invaders, and that the pacifist could fairly have a right to specialize in performing other valuable services for his compatriots. Kenneth Mellanby describes an attempt to allow pacifists to do just this during the Second World War.[19] Pacifist volunteers were used at the Sorby Research Institute to investigate various medical problems for which animals did not offer suitable 'models'. These experiments were unpleasant and uncomfortable, rather than dangerous, but Mellanby (not himself a pacifist) comments that 'The conduct of the volunteers showed them to be tougher and more willing to suffer danger than most other members of the population.' Clearly individuals who are prepared to act in this way aid the survival of their group even if they are not prepared to use violence in its defence. It seems to me that Narveson is unreasonable in the amount of importance which he attaches to contributing to defence against human destructiveness and that pacifists might fairly say that they prefer to save life by methods which do not involve performing other harms.

Medical research on animals represents a somewhat odd instance of an action which is prima facie undesirable but is defended on the grounds of a right to self-preservation since its effects are so indirect. The individual who performs research is most unlikely to make a discovery which will prolong his own life, and similarly financial support of research by lay people is unlikely to yield any identifiable benefit to a particular individual who subscribes. Indeed, it is possible that it could be

---

[19] *Human Guinea-Pigs* (London: Merlin Press, 1973).

argued that, since the harm done is certain, and the good resulting (generally) speculative, consideration of net probable physical effects alone could lead to an abolitionist stand. However, accepting that some speculative research on animals does lead to serious benefits in terms of human health, I believe that it is still not unreasonable to decide that there are moral grounds for wishing not to be associated with painful experiments. No one (I think) would attempt to argue that the lives of Third World children are less valuable than those of children in the developed countries, so there seems to be some merit in the argument of Richard Ryder that resources ought preferentially to be used to save human lives by means which do not involve contingent suffering for research animals.[20] It certainly appears that there is no reason why an individual should not make a principled decision to support famine relief rather than, say, heart research. Thus the absolute opponent of animal research appears to be able to offer the same defence as the pacifist who argues in favour of his right to choose the kind of service he offers to his community. Since resources are not infinite, he could also fairly argue that society at large should give preferential treatment to life-saving programmes which do not exploit animals, and that the world would be a better place if everyone accepted his views. (I doubt whether this sort of argument is sufficient to justify direct attempts to interfere with research which offers genuine, non-trivial benefits where no switching of resources could reasonably be expected to follow from this intervention.)

Of course, the attitudes of many anti-vivisectionists could more fairly be described as simply 'our group' (the animals) versus 'their group' (the scientists). Is absolute opposition to animal experiments justifiable in these circumstances? Whether one's loyalties should be to one's biological species or to one's social companions poses a whole series of different questions, though I think it is difficult to say that, for example, a blind person who opposes research on dogs because of gratitude to his own guide is acting wrongly. People who depend on animals for emotional support, however, tend not merely to be criticized if they suggest that they owe a debt of gratitude, but

[20] *Victims of Science*, revised edn. (London: NAVS, 1983), 58.

also to be accused of self-indulgence and unwillingness to care for other humans. However, recent studies have shown that the companionship of an animal can mean the difference between survival and total social collapse for some people.[21] It does not appear unreasonable that they should, in turn, be concerned about animal welfare.

While it may be possible to universalize the principle that one ought to be concerned about one's social partners, it does not seem that the particular case of 'animal lovers' could be a universal one. I may have a duty to be concerned about my close relatives, but this does not imply that you are equally bound to show regard for them. Some people may have particular duties towards animals which others do not. What I do think animal lovers' can claim is that they are responsible for setting out the right to our consideration of beings whose interests we should otherwise wrongfully ignore. If granting absolute rights is supererogatory, it may still be true that some rights are due to animals (and indeed to guilty people). In this sense, the stand of individual absolutists may be fully justified (because psychologically they *can* only be absolute in their beliefs and there are important moral considerations which they successfully raise).

I doubt whether individuals who oppose animal experiments because they identify with animals have any chance of success unless they embrace the principles I have outlined above, or attempt in other ways to adjust their arguments to the basic moral perceptions of the majority (i.e. a very large amount of animal suffering can be opposed on the basis that animals have value, although this is less than the value of a human being, and they have rights not to be caused to suffer for other than extremely serious reasons). There is no obvious basis upon which their power base could be expanded by converting non-concerned individuals to exactly the attitude which they hold, and it seems unlikely that they could physically force changes to which the majority are strongly opposed.

Even if most people do not, and perhaps are unlikely to, accept fully the thesis that inflicting harm is never justified, there may still be a valuable role for the absolutists in providing

[21] e.g. R. S. Anderson (ed.), *Pet Animals and Society* (London: Ballière Tindall, 1976).

a reminder that we do need to be certain that regrettable necessity really is necessity and not mere convenience. At present, wars involving the slaughter of large numbers of relatively harmless people are undertaken for reasons which frequently do not seem to add up to compelling necessity. Similarly, at least some experiments which are nominally justified on the grounds of essential medical need appear to be done for the sake of obviously non-essential benefits to humans. For example, studies of the effects of exposure to 'white smokes', which are used in military and fire-service training, done at the Ministry of Defence Laboratory at Porton Down must have involved very considerable suffering:

3 rabbits died within 6 h of exposure . . . They showed acute inflammation and necrosis of the larynx and trachea, together with alveolitis [inflammation of the air-passage endings within the lungs] and in one case petechial haemorrhages [small, speckling regions of broken vessels] in the lungs. . . . 4 more animals died within 24 h of exposure, showing similar changes in the respiratory tract. The 3 rabbits exposed to composition I, and surviving to the end of the experiment showed much less florid pulmonary changes and laryngeal or tracheal necrosis was not seen. Of the rats exposed to composition I, 4 died within 24 h and 1 became so moribund that it was sacrificed [killed]. All 5 animals had congested lungs with alveolitis, petechial haemorrhages and in 2 cases oedema. . . . The surviving rats had mild to moderate laryngeal and tracheal inflammation. . . .

Changes seen after composition II were qualitatively similar to those seen after composition I except that rabbits dying within 24 h of exposure had moderate pulmonary oedema. In the lungs of rabbits dying 5, 6 or 7 days after dosing frank bronchopneumonia was seen. No rabbits survived to 14 days. One rat died 30 min after exposure to composition II with tracheal inflammation and necrosis, . . . The other 2 rats which died before 6 h had elapsed from dosing showed congestion and laryngotracheitis. 2 more rats dying up to 24 h after exposure showed pulmonary oedema and petechial haemorrhages of the lungs. . . .

The high degree of local irritancy produced by the 2 materials concurred with previous observations on human exposures and dogs. . . . Further work is being carried out using repeated exposures to hexachloroethane zinc chloride smokes which will determine whether progressive lung damage occurs with recurrent exposure.[22]

---

[22] T. C. Marrs, W. E. Clifford, and H. F. Colgrave, 'Pathological Changes

No pain-killers were given to the animals at any time. In a Commons written answer to a question about the research at Porton, the Secretary of State for Defence informed the House that the Chemical Defence Establishment, Porton Down

is principally responsible for research and development in the field of defence against chemical attack. . . . It is also necessary to ensure protection against biological warfare. Other work includes the study of wounds so as to improve medical understanding and hence treatment. . . . Experiments must be performed 'with a view to the advancement by new discovery of physiological knowledge or of knowledge which will be useful for saving or prolonging life or alleviating suffering'. . . . In 1983 CDE undertook about 9,500 animal experiments . . . In order to limit the risk to human beings in cases of riot control work has been done on safe use of control agents such as CS and baton rounds. . . . all work at CDE, Porton Down is . . . conducted by appropriately qualified personnel to the highest standards . . . We believe the work to be of extreme importance and in the national interest.[23]

It is quite possible to say that animals have a right not to be treated like this, while agreeing that it is also true that only saints and heroes could be expected (for example) to be prepared to do without insulin preparations if they happened to suffer from diabetes.

A number of experiments are carried out to develop medicines which provide only marginal benefits over those already in use. A large number of these are psychoactive drugs which may well be actually harmful in the long term (although profitable to the firms which make them) because they encourage patients to see their personal problems as a sickness to be treated, rather than as difficulties which they are capable of resolving themselves. Experiments on animals to develop new cosmetics and toiletries, such as shampoos, have been justified on the grounds that it is essential to ensure that new products will not cause injury to human users.[24] Paton considers that our option to do without new and potentially injurious cosmetic products is not a valid reason for ceasing animal tests because

Produced by Exposure of Rabbits and Rats to Smokes from Mixtures of Hexachloroethane and Zinc Oxide', *Toxicology Letters*, 19 (1983), 247–52.

[23] *Hansard*, 22 Mar. 1984, c. 570–2.
[24] W. Paton, *Man and Mouse* (Oxford: Oxford University Press, 1984) 140–4.

it provides no protection for 'foolish and careless' people who would continue to use new products regardless of risk. (He does not explain why it would not be possible to legislate to forbid the introduction of new cosmetic substances, just as it is currently possible to prevent the sale of products which have proved harmful in animal tests.) He also argues that it would be difficult to decide which products are cosmetics and which are medicinal (for example, cream for dry skin, salve for chapped lips, shampoos). However, at least some cosmetic substances clearly are purely decorative and where any potential 'medical' benefits are so trivial as to make it difficult to decide whether a product is cosmetic or medicinal it seems not unreasonable to say that humans should be prepared to make do with what is already on offer. Just because there are some borderline cases where there may be genuine doubt about the seriousness of the human interests involved (for example, treatments to prevent, or conceal, disfigurement) does not mean that we are thereby absolved from making decisions about cases which are not borderline.

Cosmetic-testing experiments represent a small percentage of the total number of animal experiments done in Britain but the total number of animals used for this purpose is large because the overall number of animals used in experiments is so great. In 1987 a total of 14,534 experiments on animals was done to test cosmetics for safety, 6,916 to test household products, and 3,253 food additives. The total number of all experiments was 3,631,393.[25] The numbers of animals who are killed for food or kept in unsatisfactory intensive conditions runs into the hundreds of millions every year, and each year thousands of pet animals are killed because they are unwanted. Accepting even a very limited and non-absolute view of animals' rights would mean significant benefits for substantial numbers of animals.

[25] *Statistics of Scientific Procedures on Living Animals, Great Britain 1987.* Cm. 515 (London: HMSO, 1988).

9

ARE HUMANS MORAL? THE PROBLEM OF SOCIOBIOLOGY

The theory of group selection has taken most of the good will out of altruism. When altruism is conceived as the mechanism by which DNA multiplies itself through a network of relatives, spirituality becomes just one more Darwinian enabling device . . .

Scientists and humanists should consider together the possibility that the time has come for ethics to be removed temporarily from the hands of the philosophers and biologicized . . .

E. O. Wilson, *Sociobiology*.

An interdisciplinary discussion of ethics must deal with the claim made by some biologists that our belief that we can act morally is a delusion produced by our evolutionary history. As a biology undergraduate, I can remember a group of us sitting rather glumly contemplating group selection taking the goodwill out of altruism, but it is important to recognize that most biologists will not be prepared to accept a theory about the way we ought to act unless it can be shown to be compatible with known biological facts.[1] Probably the most influential recent discussions of the biological basis of morality are E. O. Wilson's two works, *Sociobiology*[2] and *On Human Nature*.[3] Similarly, B. F. Skinner says:

---

[1] For a discussion between two philosophers and a biologist on this see J. L. Mackie, 'The Law of the Jungle', *Philosophy*, 53 (Oct. 1978), 455–64; Mary Midgley, 'Gene-Juggling', *Philosophy*, 54 (Oct. 1979), 439–58; Richard Dawkins, 'In Defence of Selfish Genes', *Philosophy*, 56 (Oct. 1981), 556–73; J. L. Mackie, 'Genes and Egoism', *Philosophy*, 56 (Oct. 1981) 553–5; and Mary Midgley, 'Selfish Genes and Social Darwinism', *Philosophy*, 58 (July 1983), 365–77. See also R. D. Alexander, *The Biology of Moral Systems* (Hawthorne, NY: Aldine de Gruyter, 1987).

[2] (Cambridge, Mass.: Harvard University Press, 1975.)

[3] (Cambridge, Mass.: Harvard University Press, 1978.)

the way one person treats another is determined by reciprocal action. We gain nothing by turning to feelings. It is often said that people comfort the distressed, heal the sick, and feed the hungry because they sympathize with them or share their feelings, but it is the behavior with which such feelings are associated which should have had survival value and which is modified by countercontrol. We refrain from hurting others, not because we 'know how it feels to be hurt', but (1) because hurting other members of the species reduces the chances that the species will survive, and (2) when we have hurt others, we ourselves have been hurt.[4]

F. L. Marcuse and J. J. Pear claim that a scientific analysis of the extent to which alternative sets of ethical values will tend to promote the survival of a culture can in principle tell us what values will ultimately be incorporated by those cultures which survive.[5] Ethical value-sets which are inferior in terms of promoting survival can only ever be transitory, so that survival of the culture is the value from which all other values flow. They suggest that eventually the most successful societies will be those which design their own value-sets and install them in their culture by the use of an effective technology of human behaviour.

One serious danger in this kind of thinking seems to be the probability that designers attempting to create improved value-sets by scientific analysis of survival value would sometimes get it wrong.[6] If these mind-sets were then simply imposed on people by conditioning instead of getting the normal processes of moral evaluation to which we ordinarily subject new ideas, social change could go off in disastrous directions.

Even if this sort of technology of values could be developed it would not necessarily eliminate the need for ordinary human moral choice. Even simple models of social behaviour in animals show that there is usually not just one supremely

---

[4] *About Behaviorism* (London: Jonathan Cape, 1974), 192.

[5] 'Ethics and Animal Experimentation: Personal Views', in J. D. Keehn (ed.), *Psychopathology in Animals* (London: Academic Press, 1979), 308.

[6] See e.g. R. L. Boot, A. G. Cowling, and M. J. K. Stanworth, *Behavioural Sciences for Managers* (London: Edward Arnold, 1977). Over the years industrial managers have been provided with a series of quite different scientific recipes for success, all proffered with great confidence. This is not necessarily a bad thing if the input is seen just as one extra source of new ideas.

successful pattern, but that there are often several possibilities which are all stable. If scientific analysis offered us more than one set of possible values we would need to choose between them and this choice would be a moral one, even if we were to make the decision *not* to use our sense of moral rightness, but to flip a coin instead.

The term 'sociobiology' was introduced by Wilson to denote the scientific study of ways in which the social behaviour of animals (including humans) is shaped by biological processes. In sociobiological literature, there are two terms which crop up persistently: 'selfish' and 'altruistic', which are used by biologists in special, technical ways; rather as chemists find it helpful to talk about hydrophilic (water-loving) and hydrophobic (water-hating) molecules. No one is likely to be misled into thinking that we are compelled to take any particular attitude towards the substance H_2O because our tissues' properties depend crucially on a complex balance of hydrophilic and hydrophobic elements. However, Wilson himself occasionally tends to treat the biological definitions as though they were really improved versions of colloquial usage.

Technical definitions of 'altruism' and 'selfish' are to be found in Richard Dawkins's book *The Extended Phenotype*:

An entity, such as a baboon or a gene, is said to be altruistic if it has the effect (not purpose) of promoting the welfare of another entity, at the expense of its own welfare. Various shades of meaning of 'altruism' result from various interpretations of 'welfare' . . . *Selfish* is used in exactly the opposite sense.[7]

Such 'altruism' does not imply the existence of any of the mental states we normally associate with conscious human altruistic behaviour. In order to maintain the distinction between the two meanings, I shall here use 'altruism (I)' to signify the technical, biological usage, and 'altruism (II)' where some conscious, other-directed, helpful or friendly attitude is intended. Similarly, 'selfish (I)' = self-preserving or replicating activity; 'selfish (II)' = conscious promotion of self at the expense of others. It is logically impossible for altruism (I) to be stable at the genetic level, because the effect of an altruistic (I) allele (gene form) is simply to cause itself to be replaced by an

[7] (Oxford: Freeman, 1982), 284.

alternative form. This means that subsequent generations necessarily contain no (or at least fewer) altruistic (I) genes. Altruism (I or II) *can* exist at the level of the individual organism because behaviour which harms the individual actor can benefit copies of her genes residing in the bodies of other individuals, such as her children. (For example, a mother who dies to save ten children multiplies her genes five fold although she perishes as an individual.) Hence, any evolved organism will inevitably possess a genetic history which biases it to behave in a way which has a statistical tendency to promote replication of its own genes. It must be emphasized that this connection is only a statistical one, and that self-sacrificing behaviour does not have to be consciously calculated, or to 'get its sums right' all the time. If mothers generally save more genes than they lose by self-sacrifice, then this trait will tend to be stamped in even if some mothers die without succeeding in saving their children or die to save an adopted child who is emotionally, but not biologically theirs. 'Surface-level' behaviour need only be genetically right on average, and may, or may not, be accompanied by altruistic II feelings.

Wilson sometimes writes as though 'ethics' equals 'altruism'. It seems desirable to make clear at the outset that a proof of the evolutionary impossibility of genuinely altruistic (I or II) behaviour would not prove the impossibility of ethics, in the sense of reasoning about what ought to be done (as opposed to mere action on emotional impulse).

Two fundamental lines of thought seem to be involved in the development of 'biological' systems of ethics:

1. The idea that our behaviour is biologically determined; that we have a tendency to develop moral beliefs only because this trait had survival value for our ancestors' genes; and that human ethics have no significance apart from this survival function.

2. The idea that ultimate value resides in our genes (and perhaps also in the survival of our culture); and that the primary aim of all human action, including morality, is the preservation of our own genetic material.

Wilson, for example, can be seen to claim:

1. That present human choices (in the absence of biological

understanding) are guided exclusively by our evolved emotional tendencies:

The brain is a product of evolution. Human behaviour—like the deepest capacities for emotional response which drive and guide it—is the circuitous technique by which human genetic material has been and will be kept intact. Morality has no other demonstrable function.[8]

2. That the true aim of our lives ought to be the conscious preservation of our genes within a thriving human gene pool:

the new ethicists will want to ponder the cardinal value of the survival of human genes in the form of a common pool over generations . . . The individual is an evanescent combination of genes drawn from this pool, one whose hereditary material will soon be dissolved back into it.[9]

Wilson proposes that humankind's capacity to develop ideas of right and wrong has survival value to the genes because it may promote social co-operation, allow expulsion of incorrigible cheats, etc., and that this is therefore the 'function' of morality. He is here using the word 'function' (and similarly 'purpose' in other passages) in the biological sense of the (historical) reason why a particular behaviour has selective value, or the part a particular developmental stage plays in the life cycle of an organism. This use is fine so long as it is clear that only the biological sense is intended: the function of an adult human is reproduction, that of a pelagic (free-swimming) barnacle larva is dispersal, and so on. Unfortunately, Wilson then carries on to reinsert his definition of function into statements about normal human purposive behaviour. At this level we have to say either that the 'function' of actions is the satisfaction of needs, desires, etc. of conscious agents, or that we simply have no ultimate function or purpose. It is quite misleading to speak as though biologists have discovered that the purpose of life is gene replication.

A concise exposition of the logical and factual evidence in favour of some form of sociobiology may be found in the works of Richard Dawkins.[10] These offer a more detailed examination

---

[8] Wilson, *On Human Nature*, p. 167.

[9] Ibid., pp. 196–7.

[10] e.g. *The Selfish Gene* (Oxford: Oxford University Press, 1976) and *The Extended Phenotype*.

of common misconceptions about the evolutionary process than can be accomplished here, and I shall attempt only a very brief account of some errors which have been very important in shaping popular ideas about 'evolutionary ethics'.

In thinking about the biological basis of behaviour it is important to remember that a gene which would benefit the survival of the species cannot spread within any population unless its properties also ensure that it replicates at a faster rate than competing genes in that particular population. A real example of this effect can be seen in mountain gorillas. These apes live in small bands, composed of females, young, and (normally) a single adult male, who leads. If this male leader is killed or dies, then the band will be taken over by another male, who usually kills all unweaned infants. It is in his reproductive interest to do so because female gorillas suckle their infants for one to two years, and during this time they cannot conceive. If their infant dies, they become reproductively active once more. Hence a silverback (adult) male who takes over a new band and does not kill the infants will waste up to two years of his own reproductive life guarding young who do not bear his genes. In an area where there is a heavy mortality of silverbacks through poaching it is possible that he will not live long enough to father any children of his own. Thus infanticide increases his chance of leaving offspring. However, where gorillas are regularly hunted, males may not live long enough for their offspring to be weaned. Thus a band of females may pass from male to male, each time losing all the unweaned infants, and the species is severely threatened.[11]

Clearly for the sake of the mountain gorilla species, it would be highly advantageous if males lost their tendency to kill unrelated infants. However, at the genetic level where natural selection operates, a gene favouring altruistic (I) protection of strange infants *cannot* spread. Any male who possesses this characteristic will have even less chance than normal of leaving any surviving offspring, since his entry into the race to leave weaned progeny will necessarily be delayed.[12] This

---

[11] D. Fossey, *Gorillas in the Mist* (London: Hodder and Stoughton, 1983).

[12] For a review of infanticide in primates, see Sarah Hrdy's book *The Woman that Never Evolved* (Cambridge: Mass. and London: Harvard University Press, 1981), 76–95.

infanticidal tendency among males can coexist with devoted care for their own offspring. Sarah Hrdy's co-worker, Jim Moore, observed that one langur monkey was able to preserve her infant from destruction by transferring him from her own natal group to a bachelor troop, where he was protected by an old male who was probably the ousted leader of the group from which mother and infant originated. The old male took over the task of transporting, grooming, and warming the infant, while the mother rejoined her own group.[13]

A species-preserving gene which cannot compete on an individual level with other genes can never become widespread, simply because the timespan needed for selection at species or group level is so much greater than at individual level. Selfish (I) individually beneficial genes will always swamp species-benefiting genes before the latter can spread widely enough to have an evolutionary effect. If the altruistic (I) unsuccessful gene is essential for the survival of the species, then that species will simply become extinct. However, in this case the creative effect of natural selection of individuals is absent.

Such individual selection can generate animals with the ability to overcome some 'no-win' situations, thereby indirectly safeguarding the species. The Arnhem Zoo in Belgium possesses a large group of chimpanzees who live relatively freely in an enclosure several acres in extent. These chimpanzees were the subjects of an extended study by the ethologist Frans de Waal.[14] At the time of his observations, the group was dominated by three males, Nikkie, Yeroen, and Luit, who respectively ranked first, second, and third in the hierarchy. Yeroen, however, was initially the most successful at mating with females because he had perfected a system by which he was able to play off the other two dominant males against one another. If Luit approached a female he would appeal to Nikkie for support in driving him away; conversely, Yeroen would seek Luit's support against Nikkie. Since Yeroen held the balance of power which ensured Nikkie's continued overall leadership it appeared that Nikkie could not successfully attack Yeroen for courting females, since Yeroen would then appeal to Luit for support. After about a year of this, the situation changed: Nikkie and Luit ignored one another's sexual behaviour and by adopting a policy of mutual non-interference both were able to increase their reproductive success. Chimpanzees are unlikely to understand the probable genetic consequences of allowing cheats to exploit their social co-operation, but some descriptions of their behaviour do suggest that they feel consciously indignant at cheating acts:

chimpanzee group life is like a market in power, sex, affection, support, intolerance and hostility. The two basic rules are one good turn deserves another and 'an eye for an eye, a tooth for a tooth'.

The rules are not always obeyed and flagrant disobedience may be punished. This happened once after Puist [a senior female] had supported Luit in chasing Nikkie. When Nikkie later displayed at Puist she turned to Luit and held out her hand to him in search of support. Luit, however, did nothing to protect her against Nikkie's atttack. Immediately Puist turned on Luit, barking furiously, chased him across the enclosure and even hit him. If her fury was in fact the result of Luit's failure to help her after she had helped him, this would suggest that reciprocity among chimpanzees is governed by the same sense of moral rightness and justice as it is among humans.[15]

Incidentally, such a rudimentary sense of fair play would presumably mean that those who believe that rights-holders must be moral agents should accept chimpanzees as the kind of rational beings who can have rights. The evidence certainly suggests that social-contract ethics of rational self-interest are a possibility for biological organisms like chimpanzees and humans. The situation of the gorillas differs materially in that, if males *A* and *B* need to 'agree' that if either dies the survivor will allow the deceased's children to live, not only is this plan much more difficult to imagine in the first instance, but it is also unenforceable.

Another misconception about natural selection is that it means we all 'only act to preserve our genes'. A splendid example of this can be found in Bruce Fogle's entertaining book *Pets and Their People*: 'But if it ever came to a choice between her [his dog] and the children it's no contest. The children are my children. They are my genes and I will protect them.'[16] I am not trying to argue that Fogle is wrong to have a policy of saving his children in preference to his pets: what is a mistake

[15] *Chimpanzee Politics*, p. 207.   [16] (London: Collins, 1983), 27.

is the idea that, at the level of human mental experience, saving one's children is simply a matter of preserving one's genes. We love our children because they are our children, *not* because of the similarity between certain chemicals in their cells and in our cells. An unrelated adopted child merits just as much self-sacrificing defence. It probably is (historically) true that the reason why we are the sort of creatures who love our children is that genes for this behaviour tend to be successful in competition with 'child-neglecting' genes, but this is something entirely different and tells us nothing about what we ought to do. In fact, we can see that it cannot be the case that we are 'programmed to preserve our genes' *tout simple*. People who are carriers of deleterious genes (for example, the gene for muscular dystrophy) are willing, indeed eager to use technology to avoid passing on these particular genes.

A detailed exposition of a moral position which is based on 'biological ethics' (as it relates to the concept of animal rights) is found in James Pascoe's pamphlet on the ethics of experimental use of animals:

animal experimentation poses ethical problems because we are ourselves animals, and we have, or possibly imagine we have, empathy with other animals.

Animals, including the naked ape, are here at all because they are adapted to maximise comfort, satiety, longevity, and to minimise Discomfort, Deprivation, Danger, Damage and Death. If we increase another animal's comfort etc. we are kind. If we increase the 5Ds we are unkind. If we do it knowingly and without cause we are cruel. We teach our children to be kind, (I often wonder what would be the result if we did not) and we consider cruelty to be an evil. Cruelty to animals is also a crime. Why? . . .

We would all do well to try to answer that question for ourselves. My personal answer would be that being kind gives me pleasure, and being cruel gives me guilt feelings, but that probably stems from learned behaviour. A more fundamental reason would be that the more cruelty there is in the world the more chance of danger, damage and death for me. Both reasons are egocentric, but that is because I believe it is futile to think for other creatures be they people or animals.[17]

---

[17] *Attitudes to Experimentation on Living Animals. Science, Ethics, Law: A Personal View* (Leeds: Education Subcommittee of the Physiological Society, 1983).

Somewhat similarly, Professor Webster suggests that a justification of intensive rearing conditions for calves could be the
increase in sheer numbers of animals which will result,
irrespective of whether or not these systems cause suffering for
individual cattle:

I am no vegetarian and while I respect the right of any individual to be
a vegetarian if he or she wishes on whatever grounds he or she
chooses, I must point out the logical fallacy in the argument that man
has no right to 'exploit' animals. Since the beginning of agriculture,
man and farm animals have lived together to exploit the basic
resources of earth, sun and water in order to survive. Ruminant
animals have been particularly valuable neighbours of ours because
they can and do exist to a large extent on food which is not directly
available to man (they convert grass into meat and milk). If man were
to cease to 'exploit' cattle he would first of all have to kill them all off
except for a few living museum pieces in zoos and then he would have
to think up an economic way to cut the grass! This is a rather flippant
answer but then it is a rather silly question.[18]

These quotations appear to reflect an underlying assumption
that biological theories about our behaviour are not only
descriptive but also prescriptive, i.e. that our evolutionary
history determines what moral attitudes we *ought* to take.
They essentially argue that exploitation of animals should be
accepted as a natural event within the context of natural
ecological systems in which predatory or parasitic animals
destroy others without mercy, and our own history as a
predatory species. On this view of things the idea of rights for
animals is absurd because the only natural way for us to
behave is to live by killing members of weaker species. (It is
only fair to point out that Webster *does* believe that we have an
obligation to reduce the amount of suffering caused to any
animal as far as is economically possible, and he also thinks it
would be desirable for animals to be allowed the 'five
freedoms' of the Brambell Report, i.e. freedom to be able
without difficulty to turn round, groom, get up, lie down, and
stretch limbs.)[19]

[18] John Webster, *Calf Husbandry, Health and Welfare* (London: Granada,
1984).
[19] F. W. R. Brambell, 'Report of the Technical Committee to Enquire into
the Welfare of Animals Kept under Intensive Livestock Husbandry Systems'.
Cmnd. 2836 (London: HMSO, 1965).

The quotations perhaps also bring out a certain difficulty which many biologists seem to have about the relation between individuals and populations. Just because the way we treat cattle at present means that there is a large number of them it need not necessarily follow that it would not be preferable for there to be fewer animals who lived long and reasonably comfortable lives. Ecologists may calculate the success of a species in terms of its biomass (weight of living organisms at any time) or numbers, but this does not prove anything about what is best for the individual conscious animal. Neither does it mean that the rest of us have to accept numbers or biomass as the ultimate good. Further, considerations about populations need not necessarily determine what duties *I* (as one individual) may have towards a *particular* animal (considered as another individual, not as a mere unit of production.[20]

It is probable that the mistaken idea of genetic changes in a population as a real conflict of individuals, coupled with the equally erroneous idea of evolution as a goal-directed process 'aiming' at the generation of greater and greater intelligence, was one cause of the disquieting hostility shown by the Social Darwinists towards people of lower than average intelligence. If one believes that mankind's 'destiny' is being impeded by hordes with 'bad' genes 'swamping' the fittest and unfairly taking advantage of their good nature, it is only too easy to convince oneself that forcible preventive measures can be justified. For example,

It is always necessary to remember that nature itself is quite non-moral, and that there are many qualities which we by no means admire, which nevertheless are often regrettably effective in the struggle for life. All through the animal kingdom one of the most successful roles is that of the parasite, and there are states of human society where such a parasite as the professional beggar is as successful as anyone else. Something of the kind is unfortunately true in Britain just now. The people we are really encouraging are not those that we think we are, for a great many of the people who get good promotion are contributing less than their share to the next generation. At present the most efficient way for a man to survive in Britain is to

---

[20] The question of whether we could have a duty to promote 'natural' situations has been discussed more fully above.

be almost half-witted, completely irresponsible and spending a lot of time in prison, where his health is far better looked after than outside; on coming out with restored health he is ready to beget many further children quite promiscuously, and these 'problem children' are then beautifully cared for by the various charitable societies and agencies, until such time as they have grown old enough to carry on the good work for themselves. It is this parasitic type that is at present most favoured in our country; if nothing is done, a point will come where the parasite will kill its host by exhaustion and then of course itself perish miserably and contemptibly through having no one to support it.[21]

Clearly, Social Darwinism cannot be compatible with the idea that evolution validates conduct: for if it does then it is positively compulsory for people to free-load if they can.

Wilson is not a Social Darwinist, in the sense of believing that the 'unfit' should be prevented from breeding, but he does claim that the human gene pool represents ultimate value, and places conscious wants and needs firmly in second place. After the establishment of the primary value of the human gene pool, he says,

The search for values will then go beyond the utilitarian calculus of genetic fitness. Although natural selection has been the prime mover, it works through a cascade of decisions based on secondary values that have historically served as the enabling mechanisms for survival and reproductive success. These values are defined to a large extent by our most intense emotions . . .[22]

I do not think that this belief can be countered completely by the argument which Peter Singer uses.[23] Singer argues that sociobiologists reason illicitly from the fact that our genetic information is potentially semi-eternal whilst we as individuals live only a short while, to the value of preserving our genes. A

---

[21] Charles Galton Darwin, *The Next Million Years* (London: Rupert Hart-Davis, 1952), 93 f. (Not the great biologist, but his physicist grandson.) Social Darwinism in general appears to be founded upon an extremely crude idea of the genetics of human populations, assuming that the human gene pool undergoes uniform mixing in each generation through random mating. Of course humans certainly do not mate at random and the kind of generalized downward or upward shift in intellectual attainment which the Social Darwinists imagine would be most improbable.

[22] *On Human Nature*, p. 199.

[23] *The Expanding Circle* (Oxford: Clarendon Press, 1981), 72–81.

convinced sociobiologist could reply that evolution creates value, because it creates animals' capacity to have good and bad experiences. He could then go on to claim that we now have the ability to act directly upon the basic sources of value, circumventing the epiphenomenal value of mental experiences, and that this is why sociobiological theory should cause a revolution in our ethical thinking. Thus, the sociobiologist is reasoning validly from information about values to conclusions about values. I think a more convincing argument against these claims is to point out that evolutionary processes are not really any more importantly basic to mentalistic experiences of values than, for example, neurochemical processes in the brain. They are necessary for the existence of such values, but they do not possess value on their own account because value is characteristically a quality of mental events alone. If we want to discuss ethics we must discuss values and to do this we must discuss mental events themselves, not their physical origins, although it may be necessary to take account of these physical facts to decide what effect certain actions may have on the mental experiences of conscious beings. Evolutionary explanations cannot explain away ethical experiences any more than knowing why pain should evolve means that it is not a real evil to any creature presently experiencing it.[24] These experiences are perhaps best compared to secondary qualities, such as colours. Red is implicit in the structure of the world of physics (as light waves of a particular frequency), but is not actual until creatures evolve with the capacity to perceive it as colour. By analogy, values are to some extent objective.

We need also to recognize that evolutionary theory and sociobiology cannot tell us anything about *motives*.[25] The sociobiologists' difficulty in keeping separate the two notions of selfishness as a mental state and selfishness as a genetic phenomenon is perhaps related to a more general behaviourist distrust of subjectivity. The behaviourist defines motivation in terms of what can be externally observed, for example, withdrawal behaviour indicates that an animal has been exposed to 'punishing' stimuli. From this, it is only a short step

[24] This point was suggested to me by Jenny Teichman.
[25] Mary Midgley, *Evolution as a Religion* (London: Methuen, 1985), 122–31.

to saying that the evolutionary causes of behaviour discovered by sociobiologists are more real than the subjective states which we suppose to motivate us, because genetic success can be objectively studied. It then seems quite natural for the sociobiologist to go on to believe that, although we may think we protect our children because we love them, we *really* only do it in order that we may selfishly promote our own genes. But this is a mistake. Humans and animals may be predetermined to care more for their close relations than for strangers but it is most *un*likely that they do this in the rather calculating way which is suggested. Social animals' conscious motives are probably more similar to the spontaneous affections and dislikes we see in quite small children. This has significance for the possibility of friendly relationships between humans and animals, because sociobiological theorists sometimes give the impression that these are necessarily delusory because animals are either mere automata or else entirely selfish (II) entities.

Animals act basically selfishly because of the selective advantages of the 'selfish' gene. Even apparently unselfish behaviour ultimately serves their collective fitness and is therefore basically selfish. Truly altruistic, self-sacrificing behavior in the interests of suprafamily organizational units exists only in man.[26]

If this were true it would make such relationships impossible for us, simply because we are not really ourselves behaviouristic creatures and we do care whether we are being offered genuine affection or not. Even if an automaton or an entirely selfish entity always acted as if it felt friendship for us we would be repelled by it if we were able to examine its real thoughts, or lack of them. But we must not be misled into believing that this is what sociobiology is enabling us to do. It is not a way of reading what others really think.

Naturally predatory animals artificially reared with prey species can often learn to 'see' the second species as social partners rather than as potential meals and treat them accordingly. Dogs reared together with lambs have been used to provide protection for sheep against predator attack (the

[26] Erik Zimen, *The Wolf*, trans. Eric Mosbacher (London: Souvenir Press, 1981), 210.

bond between the dogs and sheep means that there is no need for a human shepherd to be around to give commands—the dogs naturally defend the group).[27] The dogs who proved unsuitable were rejected because they were too playful and insufficiently attentive to the sheep, not because they showed any sign of wanting to attack them, although normally dogs will readily chase and kill sheep. Twelve of the twenty-four dogs tested showed a good degree of protectiveness towards the sheep.

The psychologist Zing Yang Kuo conducted a series of experiments in which kittens, puppies, and two species of predatory birds (Asian song-thrush, masked jay thrush) were raised with prey species (kittens with rats, mice, or sparrows; puppies with birds or rats; predatory birds with small finches).[28] When reared with a rat, mouse, or bird as sole companion the kittens or puppies became highly attached to their cage-mate and showed distress at separation. Groups of kittens or puppies showed less attachment towards groups of rats, mice, or birds with whom they were kept, but were unlikely to harm them (a few cats developed a taste for new-born rats when these were born in the cage and some learned to chase and kill sparrows, apparently attracted by their fluttering). Six out of 16 group-reared cats eventually attacked rats when separated from them for four and a half months and shown other cats killing and eating rats. (In contrast, when cats were reared by rat-killing mothers and not socialized to rats or mice 85 per cent of kittens killed either rats or mice before they were 4 months old.) Cats or dogs who had been reared with prey species as their only company retained their unwillingness to attack when given a two-month period of training during which they were allowed to see adult cats catch and kill rats (from the time they were 4 months old until they were 6 months old). No cat killed the type of rat or mouse with whom she had been reared, although 4 cats reared with albino rats eventually killed young wild rats. Cats reared with rats seemed to generalize their

[27] J. S. Green and R. A. Woodruff, 'The Use of Three Breeds of Dog to Protect Rangeland Sheep from Predation', *Applied Animal Ethology*, 11(2) (Nov. 1983), 141–61 (cited in *Animal Behaviour Abstracts*, 12(2) (1984), 159.

[28] Zing Yang Kuo, *The Dynamics of Behavior Development*, Studies in Psychology (New York: Random House, 1967) (summary of Kuo's experiments in cross-rearing).

inhibition against killing to tame mice, which one would expect to be more attractive prey. Predatory thrushes raised under similar conditions with small finches would allow them to take food from their beaks and would brood them on cold nights.

These, and similar observations, show that such animals cannot in any sense be coldly calculating probabilities of genetic success. It seems far more plausible to say that they have inbuilt tendencies to be genuinely friendly towards individuals they have known for a major part of their lives, and that the *reason* for this is the fact that such individuals historically were likely to be genetically related. (I am not suggesting that altruism (II) is itself learned—so far as I know all cross-rearing experiments have been done with naturally social species. One would not expect, for example, that an irascible, solitary animal like the European robin could be trained to behave socially outside the breeding season.)

The reactions of animals considerably less intelligent than humans are a good deal less automatic than Wilson or Dawkins's theories might suggest. Lions, like gorillas, will attack and kill strange infants, but they do not *always* react predictably. George Schaller noted the reaction of one male lion to a cub who had been killed by an associated lioness: 'The male discovered the carcass, grabbed it, but instantly dropped it, having discovered what it was.'[29] This reaction is surprisingly 'human'.

Human and animal altruism (II) may well have a selfish (I) basis, but this cannot invalidate its actual present existence as Pascoe and Wilson seem to suppose it does.[30]

Strict sociobiologists do have a tendency to suspect the worst of everyone's conscious motivations as well as their genetic origins. Wilson, for example, concludes that the apparent self-sacrifice of people like Mother Teresa can be explained by their belief that good Christians will be rewarded

[29] G. B. Schaller, *The Serengeti Lion: A Study of Predator–Prey Relations* (Chicago and London: University of Chicago Press, 1972), 77.
[30] Mary Midgley makes a related point in her book *Wickedness: A Philosophical Essay* (London: Routledge & Kegan Paul, 1984), 181. Desires can be perfectly genuine and felt even though they have been shaped by evolutionary factors, such as the preservation of genes which are held in common with relatives.

in an afterlife.[31] Alternatively, he suggests that they believe missionary work furthers the interests of their Church, and that a tendency to promote one's own group is inbuilt in human beings because of its general selective advantage. Thus Mother Teresa need not *really* care about the people she helps. This, I believe, is fallacious. It can just as easily be proposed that faiths like Christianity 'work' by channelling our altruistic (II) care for relatives into a broader concern for all other people by inducing us to see them as 'family' and by encouraging a rational 'mental flip' which enables us to realize that others have feelings like our own (in Christian terms as children of one Heavenly Father, or in Buddhist terms as relations in former incarnations,[32] and so on).

People do seem to have a distinct tendency to create 'fictive' relations when they want to reinforce purely social bonds, for example, all the 'family' terms for types of Christian clergy, the 'brotherhoods' of trade unions and masons, sisterhoods of feminists, and so on. Equally, domestic animals seem to fit in quite easily as family members and are then defended as if they were children or siblings. The Scottish Society of the Horseman's Word, an association of agricultural labourers working with horses, apparently specifically included in its initiation procedure an oath to regard the horse, and fellow horsemen, as equal brothers.[33] There seems no good reason why such emotional channelling could not enable us to extend the boundaries of our concern in a rational way.

If sociobiology is interpreted rigidly it leads to a logical impasse: any emotion or behaviour which lacks a plausible biological function must be unreal; but, conversely, if it does have an evolutionary explanation, then it can have no ultimate value, nor can it serve as evidence of consciousness, mind, or altruistic (II) intent since biological automata should behave similarly. Attempts to discuss questions about the consciousness of other animals or our own ability to act morally are placed in a perfect double bind. I can see no argument (or indeed evidence) which could be expected to offer a way of escape from

[31] *On Human Nature*, p. 164 f.

[32] H. Saddhatissa, *Buddhist Ethics* (London: George Allen and Unwin, 1970), 97.

[33] George Ewart Evans, *Horse Power and Magic* (London: Faber, 1979).

this; I can only say that, whatever may be going on amongst the genes, we still are aware of the existence of a variety of conscious minds and of the need to decide what ought to be done.

It may be true, as Wilson says, that we are mere epiphenomena:[34]

The mind will be more precisely explained as an epiphenomenon of the neuronal machinery of the brain. That machinery is in turn the product of genetic evolution by natural selection acting on human populations for hundreds of thousands of years in their ancient environments.[35]

A true account of human behaviour would then be as follows. In any given situation, neurones fire, compute the probable 'best' action, with appropriate accompanying feelings as surface phenomena, and then activate the required muscle cells. This process cannot always be tightly genetically controlled, since it is an obvious fact that many animals (including humans) do possess considerable ability to modify their behaviour in accordance with experience. However, such animals would depend on underlying genetic constraints (governing brain size and power, basic 'drives', etc.) and these would necessarily have a selfish (I) basis.

This model of mental life seems at first sight likely to prove fatal to a sense of human worth. I believe it is possible to show that this is a mistake, and that consequently any less extreme version of physicalism must leave a place for the value of conscious life. We are aware through introspection that, at the level of consciousness, it feels as though we are motivated by love, fear, ambition, trust, hate, and so on. We are not normally aware of the possible genetic effects of our behaviour, and we certainly do not normally calculate this in any systematic way. Even if it is possible that consciousness cannot change the results of brain computations, it is this which gives the computations value. You, I, and the dog Fido may exist only as the surface phenomena of brain states, but, since these phenomena are an aspect of brain state, not some kind of

---

[34] Although I think there are in fact some convincing arguments against the belief that consciousness is always epiphenomenal, see above.

[35] Wilson, *On Human Nature*, p. 195.

helpless 'ghost in the machine', we really do love our children, care about starving Ethiopians, work for peace, etc. (Obviously we are also bodies, but robotlike bodily behaviour without mind could have no significance.)[36]

We have inherited minds which care about the feelings of other minds and which have direct experience of value. Increased biological knowledge can enable us to maximize value more effectively; it ought not to be used in attempts to either 'dissolve' or disregard these values. Ethical reasoning and argument remain possible. Our brains can accept reasoned arguments as input and compute whether these are to be accepted or rejected. If my arguments are good I may cause your brain to modify its activity giving you the experience of being convinced.

A sociobiological theory of the origin of morality does not mean that human moral decisions have to be seen as automatic reflexes, like the behaviour of ants. Ants do not need to be moral because the things they do are comparatively simple and do not involve the need for complex choices. All workers in one nest are equally related sisters, so there is no conflict between different routes for preserving ant genes: the genetic interest of any one ant is the same as that of any other member of the nest. This is not so for a mammal, whose behaviour may affect the stability of genes borne in her own body, in those of her children, siblings, parents, or a variety of others of different degrees of genetic relatedness; who may need to co-operate with unrelated strangers in order to survive; and who may possess the capacity to recognize behaviour which marks the presence of a shared genetic constitution. For example, it has been suggested that altruists might recognize altruistic behaviour in others and feel more favourably disposed towards them than towards obvious non-altruists, so enabling altruism to spread by co-operation between altruists.

Humans need to be moral because they need to be able to work out how to balance conflicting interests in a flexible way. (For example, they need to be able to make choices between their own immediate physical needs and desires; the interests of various related individuals; the likelihood of encroaching

[36] See also a similar argument that biological determinism does not imply fatalism in ch. 5 of Midgley, *Wickedness.*

upon the interests of unrelated strangers to the point at which conflict results; and so on.) This kind of flexibility has advantages over rigid, ant-like pre-programming in the same way that intelligent approaches to non-social problems have advantages over invariable instinctive reactions because intelligence permits rapid adjustments to changed circumstances. In the face of social or physical problems the ant always knows exactly what to do but cannot improve; a human or lion will make mistakes but can learn from them. Very intelligent, flexible species probably have few 'low-level' behaviours built in, but tend to be governed by more general goals, such as finding food, the method of doing so being handed over to individual problem-solving. A sociobiological explanation of human morality must imply that something akin to this has happened in the case of social relationships. Instead of knowing precisely how to treat anyone, normal humans have a much vaguer innate goal (moral feelings) built upon the conflicting requirements for the replication of human genes. A given gene must preserve the body it resides in, but this may conflict with the need to preserve copies of itself in other bodies; it must preserve other human bodies where these are necessary for its replication, for example, its body's mate, children, or necessary working partners; it must not cause other human bodies to be forced to try to destroy its body; and so on, in a wide-ranging complexity of adjustments.

The idea that this kind of adjustment could be achieved by mindless selection of genes might seem far-fetched, but it has been possible to test the theory by studying some animals whose genetic relatedness can be worked out in detail. Lions have a complex social structure, in which groups are usually composed of closely related females, associated with one or more males who are not related to the females, but are normally, although not always, related to one another. Behaviour such as willingness to share food seems to follow the pattern which one would expect from genetic theory. Unrelated males normally fight, but it is also common for unrelated males who do not have brothers of their own to form partnerships. This is genetically advantageous for both partners, even though it may mean that they share resources such as food and access to lionesses, because no single male is capable of getting and maintaining ownership of a pride. Thus *unselfish* sharing behaviour can be stable at the macroscopic level and genes which inhibit sharing will be unstable.

Even such apparently unpromising species as vampire bats show reciprocal food-sharing, food being given preferentially to individuals who had previously shared food with the donor bat. This reciprocal sharing of surplus will tend to increase all bats' chances of survival.[37]

Nor need successful group living be subject to quite such dismal constraints as Wilson seems to suggest, with the individual either reduced to the status of an expendable cog, or ultimately bent upon 'selfish' exploitation of other group members.[38]

So long as the altruistic impulse is so powerful, it is fortunate that it is almost mostly soft [i.e. depends upon the expectation of reciprocity]. If it were hard, history might be one great hymenopterous intrigue of nepotism and racism . . . Human beings would be eager, literally and horribly, to sacrifice themselves for their blood kin.[39]

Our societies are based on the mammalian plan: the individual strives for personal reproductive success foremost and that of his immediate kin secondarily; further grudging cooperation represents a compromise struck in order to enjoy the benefits of group membership. A rational ant . . . would find . . . the very concept of individual freedom intrinsically evil.[40]

As Anne Rasa discovered, dwarf mongooses are highly specialized for social living, almost to an extent which recalls that of the social insects. Within a band, only one male and female will reproduce, and the 'queen' develops physiological differences from the other females of the group. Yet, unlike the expendable ant worker, every member of the mongoose group is known to the others as an individual, will be vigorously defended if attacked by predators, and will be cared for if sick or injured. Rasa notes that all this can be explained in sociobiological tems. At least five co-operating adult mongooses are needed if any young are to be raised because a single pair are not able to cope with the problems of finding food, keeping

[37] Gerald S. Wilkinson, *Nature*, 308 (8 Mar. 1984), 181–4.
[38] *On Human Nature*, pp. 151–67.
[39] Ibid., p. 164.
[40] Ibid., p. 199.

watch for predators while others forage, and guarding the young. Thus it is genetically advantageous for a young adult mongoose to stay in the band and help to rear siblings, rather than to break away and attempt to rear young alone, or with a single partner. It takes nearly two years for a young mongoose to become fully competent at all the tasks necessary for the survival of the group, so it is beneficial for sick or injured adults to be groomed, protected, and helped to feed, in order that their skills should not be lost to the group. (Rasa herself is careful to point out that the mongooses are not actually aware that their behaviour has these effects, and that their proximal motivation is effectively explained by assuming that they simply feel strong bonds of affection towards other group members.) Thus dwarf mongooses are able to be both individualists and highly social, simply because each individual is important to the group.[41]

To the extent that human society is more complex, rapidly changing, and based on intelligence, than that of lions and mongooses it is only to be expected that human social relationships must be regulated in more complicated ways. If our moral sense has evolved to make this regulation possible, by setting general goals for intelligent social problem-solving, then we can see that its nature makes sense in terms of the kind of things which it should be able to do.

1. It must set us a very generalized goal of reconciling interests and must not entail any fixed actions which we will continue to do however damaging they may become. Only an ant can afford the simple-mindedness of helping any nest-sister and killing any stranger.

2. There must be scope for moral codes and values to be learned from our society, so that past experience can be handed on, but it must be possible for these to be re-evaluated and improved.

3. Because individuals cannot be expected to permit their own interests to be entirely disregarded in favour of the group, there must be scope for individuals to persuade the majority that they are being treated unfairly, and for moral codes to be amended if their arguments seem valid.

[41] Anne Rasa, *Mongoose Watch* (London: John Murray, 1984).

If this is so, it becomes clear how a sociobiological origin of ethics is compatible with the possibility of morally wrong actions. People can simply be mistaken (perhaps wilfully) about the right moral values, and they can choose to give too much weight to their own interests at the expense of those of others. There will always be some tension between the interests of different individuals, even though those individuals may be essential to one another's survival, and morality can never be so perfect that it will entirely resolve or balance these tensions. At a very simple level, it can perhaps be said that bad actions are just those actions which others justifiably resent, and that we learn to behave better because we dislike being disapproved of. This does not simply mean that whatever the majority approves is right; because human morality is complicated and requires thought it is possible for the majority to be wrong. Thus a minority which is persecuted unreasonably can justifiably feel resentment against the majority, and can produce reasoned arguments to convince the majority that this is true. An example might be the campaign to reform divorce laws. If the majority of people initially believe that strict laws against divorce are necessary to support the family, it is still possible for a minority to produce convincing evidence that those laws simply cause suffering to individuals and do not prevent the breakdown of marriages.

The problem of animal rights could fit quite neatly into this kind of schema, if it is assumed that there exists a body of people who have powerful social relationships with animals, and who therefore wish to argue for their protection; just as, for example, we find it quite reasonable that relatives of mentally handicapped people should band together to campaign for the rights of this particular disadvantaged group.

A moral sense of this kind is perfectly real (like toothache) even though it may be rooted in the stability of particular kinds of genes, and because of its social nature it cannot be said to be just a matter of individual taste. Individuals who act against the common moral values without good reason will be resented and punished, if only through experiencing unpleasant feelings of being disapproved of. Ethics is not part of the 'fabric of the world' in the sense of being something external to us which provides the grounds of our sense of obligation.[42] However, we ourselves are part of the world, and our evolved characteristics are part of its real nature. Because it is a brute fact that humans have genuine moral feelings they really feel guilt when they act unjustifiably against shared morality, but not if they believe that they have good reasons for their actions, even though others may still disapprove. Like a general ability to act intelligently this capacity tends to stabilize the gene complex which makes it possible. And, like intelligence, morality can readily be applied to situations which are different from those which first generated it. This can happen because of the independent reality of morality, just as it is possible for us to use our hands to type, or make model aeroplanes, even though they did not evolve for these purposes.

The biological origin of morality does not mean that any action which preserves one's own genes is morally justified, and the suggestion that this is so derives from a confusion of levels. Ethics belongs to the level of macroscopic events, of behaviour and feelings; gene stability to the molecular level. Because our moral feelings are real, like pains, an action which we can understand to be immoral remains immoral even if circumstances change so that it now possesses a tendency to stabilize any genes which may produce it. If, for example, it became advantageous for humans to kill other people at random this would not mean that the species would develop a new morality, but that humans would be losing their moral capacity, as they might lose a physical organ when it ceased to be needed. To develop new moral ideas we have to be convinced that these are really better than our old ones, and to be given reasons why this is true. If we are to make use of a sociobiological account of the origin of morality we must build up a theory which accounts for the genuine complexity of human moral choice. Naïve sociobiological theory which assumes that every variation in moral codes can be explained by direct selective value producing pre-programmed behaviour simply does not do the job.

---

[42] Bernard Williams, 'Ethics and the Fabric of the World', in T. Honderich (ed.), *Morality and Objectivity* (London: Routledge & Kegan Paul, 1985), 203–14.

10
THE STATUS OF ANIMALS

It seems to be the case that human relationships are affected by the kind of factors which sociobiologists talk about, but that most current sociobiological theories about ethics are unsatisfactory because they ignore important differences between human moral behaviour and the social behaviour of comparatively simple creatures, such as the insects. It is arguable that, in their most popularly stated forms, they are not even very satisfactory for describing and explaining the social life of animals of intermediate complexity, such as birds and mammals. In this final chapter, I hope to show how a theory of the sociobiology of ethics could be developed in a way which does allow for the full richness of human moral thought, and to describe its particular relevance to the way we treat other animals.

If morality is a sociobiological phenomenon, then evidence that human–animal relationships are often complex, requiring that the human partners engage in social reasoning of the kind involved in inter-human relationships, will lend weight to the idea that ethical thought has an important part to play in regulating such interactions.[1] This does not mean that animals which can never engage with humans in social relationships have no moral status: it would be reasonable to expect that (for example) repugnance against hurting should be extended beyond its original social context.

Much of the opposition to allowing other animals a place in human morality seems to stem from an implicit assumption that it is somehow abnormal for us to give their interests any consideration and that the only correct relationship between

---

[1] For an extensive bibliography on social relationships between humans and animals see Karen Miller Allen, *The Human–Animal Bond: An Annotated Bibliography* (Metuchen, NJ and London: Scarecrow Press, 1985).

human and non-human is that of exploitation.[2] In fact, friendships with animals seem to occur in virtually all human societies, and these are not only within the context of domestication. The association between ethologist Dian Fossey and her gorilla subjects is a prime example:

Contacting Group 4 one horrible, cold, rainy day, I resisted the urge to join Digit, who was huddled against the downpour and about thirty feet apart from the other animals. It had been many months since he had shown any interest in observers, and I did not want to disrupt his growing independence. . . . After a few minutes, I felt an arm around my shoulders. I looked up into Digit's warm, gentle brown eyes. He stood pensively gazing down at me before patting my head and plopping down by my side. I lay my head on Digit's lap, a position that provided welcome warmth as well as an ideal vantage point from which to observe his four-year-old neck injury. The wound was no longer draining but had left a deep one-inch scar surrounded by numerous seams spidering out in all directions along his neck.

At this time Digit was 14 years old and almost fully adult.[3]

Similarly, the Douglas-Hamiltons in their popular account of Iain Douglas-Hamilton's elephant studies:

Elephants like Virgo and Right Hook, and well over a hundred others, accepted us as harmless. But even after five years of living with them, only Virgo actually came into friendly body contact with us. . . . When Saba [the Douglas-Hamilton's first child] was three months old, . . . we met Virgo and her closest relatives one evening. I walked up to her and gave her a gardenia fruit, in a gesture of greeting. She was a trunk's length from me, took the fruit, put it in her mouth, and then moved the tip of her trunk over Saba in a figure of eight, smelling her. I wondered if she knew that Saba was my child. We both stood still for a long while, facing each other with our babies by our sides. It was a very touching moment. I feel sure that Virgo will remain a life-long friend of ours, even if we do not see each other for years.

In the postscript to their book Iain Douglas-Hamilton describes how the family returned to the reserve two years later:

I was especially curious to see whether Virgo would remember us. When I found her I got out and called to her. She stopped and turned

---

[2] For a discussion of these attitudes see James Serpell, *In the Company of Animals* (Oxford: Blackwell, 1986), 34–47.

[3] *Gorillas in the Mist* (London: Hodder and Stoughton, 1983), 199.

towards me, then slowly she came forward, extending her trunk to touch my hand.[4]

Nor need the relationship require that the human partner be in any sense 'soft', patronizing, or 'sentimental'. In her account of rehabilitating captive chimpanzees, Stella Brewer relates how she maintained the adult male's respect for her leadership:

. . . William walked to the shack. The door was unlocked; he opened it, then slammed it back. I heard an ominous crack as some of the wood splintered around the hinges. 'William, stop it or I'll thump you!' I shouted. . . . William sat—his coat semi-erect—and glared at me. I strode over to him, anxious to avoid his deliberate provocation while Tina [the dominant female] was around, yet determined not to lose face. . . . William got up and punched my foot. By this time I was finding it difficult not to react. His insolent taunting I found unbearable. When he swung round, grabbed my ankle and neatly flipped me over, I could stand no more. The next time he came, I kicked out hard and caught him in the thigh. He screamed, his arrogance flown with the wind. I walked over to him and cupped his chin in my hand. 'Now then, will you stop bugging me, you brute!' He listened, then pulled away sulkily and walked towards the edge of the gully. . . . When William returned that evening, he greeted me affectionately and though he continued to be boisterous, he had lost the insolent, provoking manner that he had shown at the beginning of the day.[5]

In spite of her 'tough' attitude towards the chimps Stella Brewer clearly had a strongly affectionate relationship with them, as shown in her reaction when one of the younger ones was believed to have been killed by a lion some days after the incident related above.

I searched the ground for signs of blood. In my aching, empty head I heard again Pooh's voice screaming fear . . . William touched my leg. He was holding out a dry leaf to me. When I didn't immediately accept it he tossed it to my feet but the leaf being light, settled closer to his own. He picked it up again and shredded it.

William, of course, William. Of all the hopes, the plans, the achievements, only William and I survived. I had failed. Pooh was to be my last failure. William I would take home to those who could

---

[4] Iain and Oria Douglas-Hamilton, *Among the Elephants* (London: Collins, 1975), 204, 270.
[5] *The Forest Dwellers* (Glasgow: Collins, 1978), 154.

look after him. Yula and Cameron would spend their lives safely in a cage. . . .

William perhaps could sense but could not comprehend my uncontrollable sobbing. He handed me small stones, a twig, and finally patted my back as I had often patted his and Pooh's. Then he moved away, striding out for camp. . . . I stumbled on a rock and fell down. William stopped and watched me get to my feet again, then strode on.

In the event Pooh was found alive and the rehabilitation work continued.[6]

Perhaps the most common example of such fellowship is the association between humans and dogs. The dog's social behaviour has become largely redirected from other canines to human beings. The depth of feeling which may be involved in this interspecies relationship is portrayed in a passage from the collected letters of the novelist T. H. White, written after his Irish setter, Brownie, had died:

I have an affectionate disposition, and as I could not take a wife I needed something to lavish it on, which was why, apart from Brownie, I was always fooling about with hawks and badgers and snakes and God knows what else. I know it is difficult to understand old maids like me, except with a kind of pitying contempt, but if anybody can it will be you. (I dont feel at all contemptible.) I loved Brownie more than any man I have ever met has loved his natural wife. We were like cats on hot bricks away from each other, and thought about each other all day, particularly in the last years. It is a queer difference between this kind of thing and getting married, that married people love each other most at first (I understand) and it fades by use and custom, but with dogs you love them most at last. They are meaningless to begin with, and if I bought a bitch puppy tomorrow she would not replace Brownie for a long time to come.

It is perhaps a matter of life or death to me .to know whether to recommence the same long trail with a new puppy, which I feel morally certain would end in both our deaths, because I would be too old to make a fresh start in 12 years time, or whether to have nothing more to do with dogs. In the latter case I might not be able to keep it up (living) and I dont know where I would put my surplus affections.[7]

[6] *The Forest Dwellers*, pp. 157–8.
[7] *The White/Garnett Letters*, ed. David Garnett (London: Jonathan Cape, 1968), 182.

Evelyn Cauwood of the Society for Companion Animal Studies recently conducted a survey on the subjective effects of the companionship of pet animals in arthritis sufferers. Among the responses she quotes is that of a cat owner who refers to her cat as 'a true, loyal, affectionate, life-long companion/friend . . . [someone who] more than made up for all the unpleasantness in my life . . . was my support, my tower of strength. . .' .[8]

Many other domestic animals are not simply passive victims of human exploitation, but act towards humans as if they were social partners, and receive social attention from humans. Even rough behaviour can indicate that animals are *not* seen as mere 'stocks and stones'. Who would waste time threatening a piece of wood? Where an animal has been encouraged to regard us as a social partner this may give him particular claims upon us, and this is the reason why 'treacherous' actions towards animals seem particularly repugnant (consider, for example, the public outcry against the sale of ex-cavalry horses for pet food).

Thus many animals evidently do satisfy the kinds of tests for rights which F. L. Marcuse and J. J. Pear consider are applicable to subnormal humans:

a highly evolved culture would accord these people moral rights (including the right not to be subjected to harmful or aversive experimentation) . . . First, not to do so would threaten other members of the culture . . . Second, many people with severely limited behavioural development have close relatives who are intimately concerned with their welfare and who are capable of taking social action to secure moral treatment for them. Third, many people with severely limited verbal development establish close personal (i.e. uniquely human [why so?]) relations with normally functioning people, who, therefore, also act to secure moral treatment for them.[9]

It might appear that a society which sees animals purely as tools must have some material advantages over one which starts to develop scruples about certain ways in which animals are used. (Marcuse and Pear claim that animals ought not to be

---

[8] E. Cauwood, in *Society for Companion Animal Studies News-sheet*, Nov. 1984, pp. 6–7.

[9] F. L. Marcuse and J. J. Pear, 'Ethics and Animal Experimentation: Personal Views', in J. D. Keehn (ed.), *Psychopathology in Animals* (London: Academic Press, 1979), 310–11.

given significant moral rights because this does not have survival value for the culture.)[10] However, I am not sure that this is necessarily valid. Since animals commonly used for domestication are social beings, their efficient use depends (*inter alia*) upon understanding and interacting with their social processes. It is perhaps easier for humans to do this in a natural way, rather than mechanically with no true social input of their own. If relations with animals were to become purely a matter of artificial manipulation, it seems to me that this would necessarily involve something very like lying and cheating. In any event, however skilful humans might become, elimination of all social outflow towards animals would necessarily rule out some of the benefits which we currently enjoy. Several studies have shown that a relationship between humans and pet animals can have positive effects upon human mental health, through the provision of 'someone to love'.[11] These benefits would no longer be available to a society which truly regarded animals as mere things. Thus I think it can be shown that it is not absurd to talk about the mutual benefits of society between humans and domestic animals. Dogs have been part of human society for almost as long as our own species, *Homo sapiens*, has existed.[12] Sheep and goats are only slightly more recent on the scene. The relationship must originally have been very close: as pointed out by Ryder, the young animals originally reared must have been suckled by women[13] (as is still the practice in some primitive societies),[14] since it would be essential for them to be imprinted upon humans from a very early age if the required degree of tameness was to be produced.[15] Before milk-producing animals were domesticated, human milk was the only suitable food

[10] 'Ethics and Animal Experimentation', p. 311.

[11] e.g. Leo K. Bustad, *Animals, Aging and the Aged* (Minneapolis, Minn.: University of Minneapolis Press, 1980).

[12] J. Clutton-Brock 'Dog', in I. L. Mason (ed.), *Evolution of Domesticated Animals* (New York: Longman, 1984), 198–210.

[13] M. L. Ryder, 'Sheep', in Mason (ed.), *Evolution of Domesticated Animals*, pp. 63–85.

[14] Clutton-Brock, 'Dog', pp. 208–9.

[15] But in the case of sheep see Valerius Geist, *Mountain Sheep* (Chicago and London: University of Chicago Press, 1971), 41–5, who found that American wild sheep could be readily tamed by feeding them with salt so long as they were not frightened by hunting.

available for raising very young wolf pups or lambs away from their own mothers. If women were the main agents in forming the original bonds between humans and domestic animals there is possibly a certain fitness in the present predominance of women in groups working for reform of the relationship. One might perhaps speculate that the original importance of domestic animals for human survival, and the critical role of women, is partly responsible for the apparently superior performance of women in tasks which involve relating to animals, for example, the rearing of calves (not an example of the trait in a very benign form),[16] show-jumping,[17] and the relatively greater success of women in ethology, compared with other sciences.[18]

T. Ingold suggests that there is a basic difference between the perceived status of animals kept for work and those kept only for meat production, and that this difference in attitude is virtually universal in all human societies where domestic animals exist.[19] Animals which perform physiological work, such as the production of milk and eggs, seem to be classified similarly to those who do mechanical labour, like ploughing, and both have a kind of 'sub-proletarian' status. They are seen as quasi-persons, with a definite membership of human society, and their usefulness as workers depends upon personal relationships with individual humans. Ingold suggests that this status is more or less essential for the use of work animals in a primitive society, because such animals are of no practical use unless they are fully tame and co-operative. Meat animals typically do not have relationships of taming with humans and do not have individual names. They do not co-operate with

[16] Universities Federation for Animal Welfare, *Stockmanship on the Farm* (Potters Bar: UFAW, 1983); J. Webster, *Calf Husbandry, Health and Welfare* (London: Granada, 1984).

[17] Performance appears equal in terms of successes, but male riders receive an advantage under FEI (Fédération Équestre Internationale) and BSJA (British Show Jumping Association) rules because all riders have their total weight made up to an average male value of 12 stone (168 pounds) (R. S. Summerhays, *Summerhays' Encyclopaedia for Horsemen*, 6th edn., revised by Stella A. Walker (London: Frederick Warne, 1975), 364).

[18] Possibly no other field of scientific endeavour has a comparable degree of female success: Dian Fossey, Jane Goodall, Sarah Hrdy, and Felicity Huntingford are all examples.

[19] *Hunters, Pastoralists and Ranchers* (Cambridge: Cambridge University Press, 1980).

humans but are controlled purely by coercive means and have personal relationships only with one another. He speculates that domestic herds of meat animals arose by the depersonalization of domesticated working animals in times of famine, pointing out that, in reindeer-hunting societies which possess tame working reindeer, these tame animals are eaten only when food is desperately scarce. Reindeer-dependent societies like the Eskimos which lack domestic reindeer depersonalize and consume low-status human members of their domestic group (women and children) in similar times of crisis. Where transition to full-time raising of herds for meat has occurred, as among some carnivorous reindeer pastoralist tribes, domestic reindeer remain permanently depersonalized and may be eaten at any time.

the necessary precondition for a transition to pastoralism lies in a particular characteristic of higher animals. This is their capability to *act*, in ways conditioned as much by their social environment as by genetic inheritance. In this capability resides the potential for animals to be *tamed* by man: that is, to enter into social relations of domination defined by man's subjugation of the animal's will to suit his own purposes . . . Marx, in fact, denied the possibility of this form of relationship between man and animals on the grounds that animals lack will: 'Appropriation [of another's will] can create no such relation to animals . . . even though the animal serves its master . . . Beings without will, such as animals, may indeed render services, but their owner is not thereby lord and master' [Karl Marx, *Pre-capitalist Economic Formations*, ed. E. J. Hobsbawm (London: Lawrence and Wishart, 1964)]. Similarly in *Capital*, domestic animals are classified alongside primitive tools as *instruments* of labour . . . This, however, is to relegate animals to the status of mindless machines. In truth, the domestic animal is no more the physical conductor of its master's activity than is the slave: both constitute labour itself rather than its instruments, and are therefore bound by social relations of production. In other words, taming is not a technological phenomenon.[20]

Again, considering the Chukchi, reindeer pastoralists inhabiting the Asiatic side of the Bering straits, Ingold speculates:

at the risk of venturing into the absurd, I would like to suggest that there does exist a kind of class exploitation, not within Chukchi

---

[20] Ingold, *Hunters, Pastoralists and Ranchers*, p. 88.

society, but in the relation beween the Chukchi and their herds. For if the 'surplus' that accrues to the primitive accumulator is of live stock, its formation depends ultimately on the productive labour of the animals themselves [searching out and consuming plant food], rather than on the appropriative labour of their herdsman [who merely *prevents* animals being consumed by human or animal predators other than his own group]. Consider, for the sake of comparison, the relation of slavery or serfdom, in which 'one part of society is treated by another as the mere *inorganic* and *natural* condition of its own reproduction . . . *Labour* itself, both in the form of the slave as of the serf, . . . is placed among other living things *as inorganic condition* of production, alongside the cattle or as an appendage of the soil' [Karl Marx, *Pre-capitalist Economic Formations*, ed. E. J. Hobsbawm (London: Lawrence and Wishart, 1964)]. My point is that precisely the same may be said of the herds of pastoralists: quite simply, one *species* is treated by another as the natural condition of its own reproduction. This much is suggested by Marx's inclusion of slaves and serfs 'alongside cattle'. In a society with herds but without slaves or serfs, class and species boundaries coincide.[21]

Some people may feel that it is a very bad thing for animals to be slaves, serfs, or even proletarians, but this status actually represents a considerable improvement upon that of mere objects or 'tools'. Societies in which humans were kept as slaves on the whole did not consider that such people had no moral status at all: *eating* slaves or killing them for trivial reasons could still be considered to be morally and even legally wrong.

Another anthropologist, Roy Rappaport, who studied the Tsembaga Maring in New Guinea notes:

It may be suggested that the petting and stroking to which Maring pigs are subjected as infants is an additional factor [besides human provisioning] in keeping them domesticated throughout their lives. Such handling by humans communicates and produces positive affect, through which . . . the pig is bound to a social group dominated by humans. It is hardly facetious to say that the pig through its early socialization becomes a member of a Maring family.[22]

Rappaport records that young pigs are treated as pets, share the living areas of the woman's house until they are 8 months to 1

[21] *Hunters, Pastoralists and Ranchers*, p. 234.
[22] *Pigs for the Ancestors* (New Haven: Yale University Press, 1967), 59.

year old and thereafter are given individual stalls inside the house, to which they return at dusk from foraging in the forest. Until 4 or 5 months of age (when they are considered old enough to go in the forest alone) young pigs accompany the women to their gardens, initially on the leash, later following like dogs (p. 58). Pigs are given individual names (p. 60) and, although pigs *are* killed and eaten, this is only permitted in the context of religious sacrifice (p. 3) and individual pigs may in fact live out their natural span (since there is a twelve-to fifteen-year gap between 'pig festivals' and outside these festivals only a few pigs are sacrificed, mainly in response to serious human misfortunes, such as severe illness). The time lapse between 'pig festivals' appears to be determined by the point at which growth of the pig herd threatens to become deleterious to human well-being: because care of pigs requires an unacceptable amount of labour by the women; insufficient food is available for them; and they begin to do an unacceptable quantity of damage to the village gardens (pp. 156–62). Pig numbers are limited by castration of all male animals (the females are impregnated by wild boars), and the growth of the pig herd to this threatening number is relatively slow (p. 70), in contrast to Western practices, which aim to maximize rate of breeding (and hence rate of slaughter) of domestic animals kept for meat. Maring pigs seem not to be kept only for meat, but also as garbage disposers, as cultivators (ploughing up secondary growth in gardens by rooting), and as pets (p. 154).

Again, a third anthropologist, Evans-Pritchard, notes in his book on the religion of the Nuer people:

People . . . find it hard to comfort a youth who has lost his *dil thak*, his favourite ox, for he is young and the ox was perhaps his only ox and has been his companion. He has cared for it and played with it and danced and sung to it. He now sits by himself and pines, and his friends try to cheer him up. They tell him that he must not be tearful or God will be angry: 'God is good, he might have taken you, but he has taken your ox instead.'[23]

---

[23] E. E. Evans-Pritchard, *Nuer Religion* (Oxford: Clarendon Press, 1956), 13–14.

Evans-Pritchard also notes that a local missionary reported to him that she had frequently been asked by the Nuer whether she thought that cattle had souls.[24]

Ingold states that peoples whose work animals are of a different species from those they consume for food typically see the working species as group members to the extent of refusing to eat them even when killed or disabled by accident, disease, or age. Most horse-culture American Indians did not eat naturally killed horses except under famine conditions, and the British are notoriously reluctant to consume horse or dog flesh. A taboo against consumption of dogs is common in many societies. Conversely, societies which ranch animals solely for their meat tend to generate a situation of mutual hostility and antagonism, in which the cruelty of man is matched by the ferocity of beast: as on the cattle ranches of South America. Ingold notes that ranchers, in contrast to pastoralists, make no attempt to guide or control their herds for most of the year. Instead, their animals are managed essentially as though they were wild game and must seek out food, water, and shelter for themselves. Pastoralists, even carnivorous ones, have a more definite tie with their herds and make deliberate efforts to move them to favourable areas of grazing. The only distinction between ranching and true hunting is a legal one, in the minds of the humans involved. Wild animals are the property of no one, or perhaps of anyone who owns the land on which they happen to be. Ranched animals have human owners with legal rights over them and their offspring.[25]

[24] *Nuer Religion*, p. 157. However, the relationship between the Nuer and their cattle is by no means idyllic in all respects: when cattle *are* sacrificed to appease the spirits the methods of killing used on some occsions (e.g. killing by suffocation to appease the python-spirit) may involve considerable suffering for the victims.

[25] It is possible that the social behaviour of domestic animals provides the germ of a solution to the vexed question of biological determination of human behaviour. A bovine 'ethnologist' surveying the diverse life-styles of cattle in an English dairy herd, South-American ranch, Indian village, or Nuer settlement, might conclude that the social behaviour of *Bos taurus* is indefinitely plastic to the effects of social environment. Comparison with the equally varied, but subtly different life patterns of reindeer in diverse social settings, however, would demonstrate that, while the number of possibilities is large, or perhaps even infinite (there is an infinite number of points on a line, even though its length is finite), its range is constrained by the basic

Since respect for the rights of animals appears to be closely tied to experience and appreciation of their natures, I suspect that campaigning for rights within an integrated society of beast and humankind is likely to prove more fruitful than a 'separate-development' policy like that of Rodman, in which humans and animals are rigorously isolated from interference in one another's lives. Such a policy, at its most extreme, leads to the despairing comment of one of my friends, a person who has devoted his life to the cause of animal welfare: 'I saw a documentary on Passenger Pigeons [an extinct species of bird] and I thought to myself, "You lucky buggers, you can't be made to suffer".' This *is* a policy of despair, not of ultimate purity, and its underlying assumptions are essentially no less arrogant and 'speciesist' than claims that humans have an absolute right to make use of other animals. Recognition that humans and domestic animals can form a true community offers the possibility of a reformed relationship instead of the assumption that only wild animals can have lives of real value, while the tame can be best served by their humane extermination through birth control. No doubt working animals are frequently exploited (although at least in the affluent West horse riders are traditionally taught to see that their mounts are fed, watered, and groomed before considering their own comfort); however, where peasants and their plough teams live in mutual poverty it is difficult to say that these animals are really being exploited by their immediate owners. Fairer conditions for both masters and animals might be a better

biological natures of the species involved. An example suggested by Ingold (*Hunters, Pastoralists and Ranchers*, pp. 139–40) is the lack of purely carnivorous pastoralism using cattle. The original wild ancestor of domestic cows lived naturally in small groups, and tended to scatter and fight when threatened by men or dogs. Hence the pastoral use of cattle (defined as nomadic herding on open grassland) was possible only with relatively small herds of tame milch animals who accepted humans as group members. Reindeer, on the other hand, bunch into large herds and flee when threatened, and can be driven and herded by men and dogs. No relationship of taming is necessary and a small group of humans is able to control the relatively large numbers of animals needed for carnivorous pastoralism, which is less efficient in terms of utilization of primary plant sources than milking of the herds. The differences of behaviour shown *within* species may be entirely due to differences in individual social experience: my cats are perfectly ordinary domesticated felines, although their parents lived as wild animals in a feral colony and were virtually unapproachable.

option than insistence that animals must be replaced by machines.

It is interesting that India, with its huge population of working bullocks, has a long tradition of co-operative effort to maintain cattle who are past useful work.[26] At present the Indian government supports around 100 *gosadans*, institutions set up to maintain 'old and infirm' cattle, and provides some additional state assistance to private *goshalas* and *pinjrapoles*, animal sanctuaries which combine care for uneconomic cattle with a certain amount of milk production. One of Lodrick's informants expressed the purpose of such schemes:

One does not . . . kill one's mother once her useful life is over and she can no longer contribute anything to the household. Even so with Mother Cow—she gives us life through her milk and is a tender and faithful companion. She deserves better than death once her useful economic life is over, and so she should be cared for and fed in her old age.[27]

Even modern dairying institutions attached to colleges and universities and run by the Indian Department of Animal Husbandry normally maintain animals until the end of their natural life span. Lodrick notes that such government activity has not even come close to solving the Indian problem of the combination of scarcity of fodder and large numbers of uneconomic cattle, but that at least some of the most efficiently run cow protection institutions were able to combine a satisfactory profit with a policy of 'pensioning off' old animals. When the cattle finally do succumb to old age or illness their carcasses are utilized for leather, meat, and bone-meal and tallow.[28] In fact Lodrick's description of Indian treatment of animals makes good psychological sense if we assume that the people he interviewed behaved with the same kind of mix of virtue, self-interest, and self-deception that we find in our own treatment of other humans. There is, for example, a striking parallel between the Indian custom of

[26] D. O. Lodrick, *Sacred Cows, Sacred Places* (Berkeley and London: University of California Press, 1981), 14–15.
[27] *Sacred Cows, Sacred Places*, p. 24.
[28] Ibid., pp. 25–8, 186–97.

placing unwanted male goat kids in *pinjrapoles*, in order to avoid direct responsibility for killing them, and Western societies' treatment of severely handicapped human babies. The kids' original owners may make a small donation to the sanctuary towards the cost of the animals' upkeep, and the sanctuaries attempt to provide adequate rations of milk, but overcrowding, malnutrition, and disease mean that there is a high mortality of such unwanted animals.[29] When Helga Kuhse and Peter Singer investigated the attitude of doctors to treatment of severely handicapped babies they discovered that substantial numbers felt that it was acceptable to allow such children to die by withholding treatment (such as antibiotics to treat infections), or even to sedate them so that they would not feed, but that direct killing (as by painless lethal injection of barbiturates) was morally unacceptable as well as illegal even though it might cause less distress to the children concerned.[30] (Kuhse and Singer reject this, and argue that infants who are so severely handicapped that their lives are likely to contain nothing but suffering should be killed painlessly.) Similarly, in Tolstoy's *Resurrection*, 'This unmarried woman had a baby every year, . . . each one of these undesired babies, after being carefully baptized, was neglected by its mother, whom it hindered at her work, and was left to starve.'[31] (Louise Maude notes that the mortality rate in one large Foundling Hospital in Moscow at that time was 9 out of every 10 children admitted (p. 245).) We do not see this particular hypocrisy as essentially 'religious' or 'superstitious' in nature, so I do not see why we should say that the Indian peasant is merely superstitious in his reluctance to take the life of an animal.

A correct sociobiological theory of human behaviour cannot suppose that humans behave simply as calculating survival machines. Human behaviour is best explained by supposing that it is subject to sociobiological constraints, but that it is mediated by consciousness, which also possesses certain

[29] *Sacred Cows, Sacred Places*, pp. 96–7.
[30] H. Kuhse and P. Singer, *Should the Baby Live?* (Oxford: Oxford University Press, 1985).
[31] Trans. Louise Maude (London: Grant Richards, 1903), 5.

constraints and characteristics essential to its nature. In particular, we should note similarities to the observation in Chapter 3 that mourning behaviour in geese can be explained by the theory that close bonds between mated pairs are beneficial to the survival of geese, and that it is because such bonding must be mediated by consciousness, not mechanical calculation, that geese also show positively deleterious mourning behaviour when a partner is lost. Evolutionary theory predicts that the benefits must outweigh the costs, and that affectionate bonding must have an overall tendency to promote genetic success but, if such bonding can be achieved only as part of a 'package' with some non-beneficial effects, then the package will be preserved by natural selection.

Examination of the richness and diversity of interactions between humans and animals illustrates that the complexity of theories of the relationships must be at least as great as that of theories about inter-human relationships. It may also suggest why people are motivated to defend the concept of 'animal rights'. There is evidence (see above) that social deprivation has profoundly damaging effects upon the psyche of intelligent social animals. (For example, chimpanzees reared in isolation treat their mirror images as mere objects, not as reflections of themselves, isolated monkeys self-mutilate; humans who suffer from autism and cannot recognize that other people have minds appear to us to be distressed.) It seems that the social contact needed to prevent these effects does not need to come from a member of the same species—animals reared by humans do not show any of the bizarre effects of rearing in isolation, although they often appear to come to believe that they are themselves human. This explains why we should have such powerful desires to protect social companions, even when they are too small, too weak, or too distantly related to offer any direct physical or genetic benefits to our 'selfish genes'. Because we are the kind of biological entity which operates using intelligent consciousness, the value to us of other consciousnesses who provide a social medium in which our consciousness can survive is much greater than their bodies' temporary value as raw materials. So it is only to be expected that defence of companions should be one of our basic drives, not something secondary which is to be dealt with after we have succeeded in satisfying the desire to reproduce. Survival is prior to reproduction.[32]

Social bonds with companion animals probably have generalized benefits for humans because of the extension of varieties of conscious experience which results from the expanded possibiities of social interaction. Pet-keeping is definitely not confined to people like T. H. White who find it difficult to relate to other humans, and the popular idea that pets are 'good for children' seems to be founded in fact.[33] The development of a social bond between human and animal takes time. In general, humans will be unable to experience the full benefits of such relationships, whether psychological or purely practical, if they are too willing to discard working or companion animals as soon as any difficulties arise, or for the sake of merely short-term benefits. Animals who play a part in human society have an individual value, just as the members of a mongoose band have value for one another. Human beings are under no obligation to try to transform themselves into perfectly efficient survival machines in opposition to their true natures, and it is likely that this would be a hopeless project in any case. It is natural and right that, for example, the bonds of affection between couples survive long after the time at which this can promote genetic success. If human moral feelings have a generalized sociobiological value, we are under no obligation to restrict them so that they will *only* serve to maximize genetic fitness. Thus our generalized intuitions that unpleasant sensations are an evil for any creature who experiences them; that conscious beings avoid death; that we are all ultimately in the same boat; and that certain kinds of conduct towards

[32] See H. B. Barlow ('Nature's Joke: A Conjecture on the Biological Role of Consciousness', in B. D. Josephson and V. S. Ramachandran (eds.), *Consciousness and the Physical World* (Oxford: Pergamon Press, 1980), 81–94), who claims that consciousness depends upon interaction with other conscious individuals.

[33] J. C. Filiatre, J. L. Millot, and H. Montagner, 'New Findings on Communication Behaviour between the Young Child and His Pet Dog', in *The Human–Pet Relationship* (Vienna: Institute for Interdisciplinary Research on the Human–Pet Relationship, 1985), 50–7; G. Guttman, M. Predovic, and M. Zemanek, 'The Influence of Pet Ownership on Non-Verbal Communication and Social Competence in Children', in *The Human–Pet Relationship*, pp. 58–63; James Stuart Hutton, 'A Study of Companion Animals in Foster Families: Perceptions of Therapeutic Values', in *The Human–Pet Relationship*, pp. 64–70.

animals are simply unfair, are all entirely valid and natural. Regard for animals should be seen, not as some peculiar, artifical invention of decadent modern society, but as a natural part of the human moral sense.

The precise formulation of our duties to animals is not specified by this theory of their origin, but this is only to be expected, since the general sociobiological theory of the origin of morality predicts that ethics should be a matter of debate, modification, and competing ideas. (See the previous chapter.) The theory is not a first-order specification of what ought to be done, but rather a second-order theory about the possibility of moral thinking. It may, however, have some implications for our ideas about what could constitute a correct first-order theory of ethics. For example, a sociobiological theory of the origin of ethics will tend to lend support to the idea that our ethical intuitions are important and significant, even where there may appear to be rational arguments against them. If we have a powerful feeling that certain kinds of action are wrong, then the idea that this feeling is likely to exist because it has been important during evolution gives some support to the validity of the intuition. This does not mean that all intuitions about ethics should be accepted uncritically: the theory claims that we have ethics instead of instincts because we need to think rationally about our actions. However, it suggests that any strong moral intuition exists for some good reason, although, because sociobiology derives from evolutionary history, it is possible that the reason is no longer applicable. Any rational first-order theory of ethics should require that such reasons should at least be examined, though they may be rejected if there are better reasons for acting against intuition.

Secondly, a sociobiological theory suggests that the reasons for adopting an essentially consequentialist first-order theory are not quite as strong as some other projectionist or quasi-realistic second-order theories might imply. For example, a theory which claims that ethics are invented by rational beings as a means to allow them to live together in harmony implies that any correct first-order theory should aim to maximize benefits to the contracting agents, while avoiding the extremes of allowable harms to minorities permitted by pure utilitarianism. A sociobiological theory claims that rational agents are not directly involved in deciding the basic structure of ethics. This means that it is possible for some kinds of behaviour to be wrong even if there is no obvious rational case against them. (For example, some kinds of deceit might now appear to be harmless in purely practical terms, but could still be wrong because we feel that they are wrong, and practising them is harmful to our self-respect.)

Thirdly, it suggests that human morality is quasi-realistic, that is, that it does not exist apart from human consciousness, but that it is not some kind of illusion or mistake.

Fourthly, the theory suggests that no single first-order theory is likely to encompass the full complexity of moral reasoning, and that there will be a variety of first-order reasons for obligations (such as promises, effects on the calculus of utility, well-founded moral rules, respect for autonomy, personal relationships, and so on).

We can consider how various first-order theories might be adapted to deal with the position of animals. For example, our relations with animals might be said to bear some resemblance to a kind of social contract. It is probable that social-contract ethics is most usefully seen as a method for determining what a rational, ethical being ought to do, rather than as an explanation of how moral obligations can compel our adherence. Any formulation which attempts the latter course seems doomed to yield conclusions which our ethical intuitions recognize as mere prudence (calculation of the benefits of reciprocal, binding promises between powerful agents), and to exclude individuals such as human children whom we clearly do recognize as true objects of moral concern. Indeed, drawing further upon the analogy with human children, it might be said that, while it is obviously impossible for animals to have full rights of participation in the process of weighing of interests, it is possible that they have limited rights of participation: for example, a right that competent adult humans make some effort to discover, and take account of, their preferences and point of view.[34] The idea of a Rawlsian original contracting position[35] might provide a method for ethical decision-making

[34] C. A. Wringe, *Children's Rights* (London: Routledge & Kegan Paul, 1981), 118–29.

[35] J. Rawls, *A Theory of Justice* (Oxford: Oxford University Press, 1972).

which could help to achieve this by avoiding the opposing extremes of egoism and utilitarianism. By adopting the veil of ignorance we can make impartial decisions about the kinds of actions which the original contractors would consider justifiable behaviour for moral agents.

The original position as described by Rawls requires that the original contractors must be all moral agents (otherwise they could not make the decisions needed for the contract), and it is assumed that they will remain so in the real world. I am not sure that this second rule is essential for the idea of a social contract. By analogy with the idea of 'living wills', in which people leave directions about the way they want to be treated if they subsequently lose the ability to make rational decisions, it seems possible that moral agents in the original position might want to safeguard themselves against the possibility of becoming moral patients[36] in the real world. If the contractors do not know whether they may be moral agents or patients later on, they cannot rationally be prepared either to deny the basic rights of moral patients or to commit moral agents to an intolerable burden of ceaseless labour for the welfare of the former. By definition moral patients are not capable of understanding the nature of moral duties so the contractors cannot reasonably decree that the right to basically fair treatment should depend upon acceptance of reciprocal obligation since this would mean giving up their own rights should they happen to be, or to become, moral patients.

One argument against the Rawlsian original position as a decision-making theory is that its requirement of ignorance on the part of the contractors reduces them so much to mere abstractions that it is difficult to see why they should choose anything at all. Including a requirement that they should not know what species they will belong to would clearly make this problem even greater. However, it is important to remember that we are attempting to consider biologically evolved, conscious entities. Because of the constraints imposed by the evolutionary process we can know that all these entities will be subject to the same needs and interests at the most basic levels: the avoidance of pain and death for themselves and for

---

[36] Subjects of moral attention who are themselves not capable of moral decisions.

their social partners, and the pursuit of pleasant experiences. Clearly, these needs will take different forms for different species, but no species is likely to have requirements which seem totally senseless to us. (For example, worker bees die when they defend their hives by stinging mammalian predators. In their case the balance of interests is clearly shifted away from individual self-preservation towards defence of the reproductive community, but we can understand why this is so, and there is no reason to think that bees go out of their way to find mammals to sting.) Similarly, the extent to which an entity is self-conscious will affect the extent to which he knowingly avoids death, but, again, this will happen in a predictable way.

In a real-world situation of more than moderate scarcity (more extreme than the situation postulated in Rawls's arguments),[37] it is reasonable that contractors should accept that all individuals have a right to give their own interests some degree of priority when there is a conflict of serious interests (generally threats to life or health), and that there should be some degree of probabilistic maximizing of benefits.

An alternative formulation of the idea of the original position, which avoids the difficulty of assuming that the original contractors could be moral patients, would be the suggestion that the contractors are all moral agents, but that they do not know whether they will have strong attachments to any moral patient in the real world. If this is the case, it will be in their interests to safeguard the interests of moral patients to avoid suffering emotional distress by mistreatment of moral patients to whom they are attached.[38]

Consequentialism can clearly be applicable to human–animal relations, with little or no modification of classical theories. Similarly, deontological formulations which propose that certain rules of conduct simply are good can readily be adapted for the purpose, and are compatible with a generalized sociobiological theory of morality, since it may be assumed that the source of such rules is evolved human nature.

It appears that some hunter-gatherer societies did indeed believe that there were implicit agreements between humans and animals. The historian Calvin Martin presents evidence

---

[37] Rawls, *A Theory of Justice*, p. 172 f.
[38] This modification was suggested to me by Jenny Teichman.

for American Indian belief in a 'contract' between humans and animals.[39] Cree and Ojibwa Indians, for example, believed that, on the one hand, human hunters had the capability to kill more animals than they needed, while on the other, wildlife, or wildlife 'keepers', the spirit guardians of wild animals, could punish humans by causing disease or by withdrawing so that the humans would starve. It was the responsibility of the Indian hunter not to kill more than he needed for survival, not to kill for fun or self-aggrandizement, and not to torture any animal. In return, the animals and their 'keepers' would refrain from taking vengeance.[40] Martin believes that this relationship between American Indians and wildlife 'broke down' when the Indians mistakenly concluded that the strange and terrifying new diseases brought by white traders were an unprovoked conspiracy of the animals against human beings. Without this restraint, the Indians caused their own downfall by over-hunting and trapping for the fur trade until they were indeed facing starvation from scarcity of game animals. He quotes the French priest Chrestien Le Clercq (1691), who noted in his account of life amongst the Micmac Indians:

> . . . the Indians say that the Beavers have sense, and form a separate nation; and they say they would cease to make war upon these animals if these would speak, howsoever little, in order that they might learn whether the Beavers are among their friends or their enemies.[41]

We may not believe in any analogous capacity for animals to enforce a social contract with us, and with Tom Regan may feel that any such enforced contract would not represent a very real respect for nature on our part.[42] (I think it is possible that human psychology encourages us to feel more real respect for those who can 'bite back' than for recipients of moral charity.) However, the Indians' 'contract' does seem to contain the basic idea that there are some kinds of action which a rational being

---

[39] *Keepers of the Game* (Berkeley and London: University of California Press, 1978).

[40] *Keepers of the Game*, pp. 128–9.

[41] Father Chrestien Le Clercq, *New Relation of Gaspesia*, trans. William F. Ganong (Toronto: Champlain Society, 1910), 421 (trans. pp. 276–7).

[42] *All that Dwell Therein* (Berkeley and London: University of California Press, 1982), 206–39.

could accept as justified even against his own interests (which is not to say that he would not still do his best to avoid becoming their object), and that there are other actions, such as torture or infliction of harm for trivial reasons, which a rational victim could never be persuaded to consider justifiable. In a sense the Indians' 'bad conscience' about killing animals which they recognized as living, feeling creatures like themselves could be said to have become externalized as the mythical 'keepers of the game'. (Compare the somewhat similar anthropological hypothesis that, in societies which oppress women, men tend to have paranoid fantasies about women getting their own back, for example, through witchcraft. The subconscious male recognition that women are being treated in a way which would justify any rational agent making an attempt to get her own back is externalized.)

The Nuer herders of the Sudan believe that all injustice, whether to humans or animals,[43] is avenged by God, who may employ lesser beings as his agents.[44] If we believe that a supposed *sanction* against injustice is purely imaginary, this does not necessarily mean we have to say that the sense of injustice itself is not a real moral belief.

One might suppose that it makes sense to say that, in desperate situations, individuals have a kind of 'veto': the point beyond which no rational choice-maker could be prepared to accept the risk of being the victim. This suggests one possible solution to the conflict of interests involved in the scientific use of animals. Some present-day experiments involve so much pain and suffering for their subjects[45] that no one could suppose that a 'rational' guinea-pig or monkey could be willing to accept them. On the other hand, not all experiments involve pain or serious harms and *some* could be regulated in a way which preserves a degree of balance between human and animal interests. It was argued earlier that the

---

[43] Evans-Pritchard specifically states that the Nuer believe that, if oxen are killed merely for the pleasure of eating meat, their ghosts will be permitted by God to take vengeance on their slayers because the oxen have right on their side (*Nuer Religion*, p. 265).

[44] Evans-Pritchard, *Nuer Religion*, p. 165 ff.

[45] Royal Society for the Prevention of Cruelty to Animals, *Pain and Suffering in Experimental Animals in the United Kingdom* (Horsham: RSPCA, 1985).

greatest prospects for 'alternative' research lie not so much in eliminating experiments on whole animals entirely, as in reducing their numbers and in preparing the groundwork of knowledge in a way which enables risks and costs to animal subjects to be decreased. For example, if it were possible to reduce the risk of serious injury or death in the final testing stage of drug assessment to something similar to the risk of driving to work in a busy city, this would be very much more acceptable than the present 'test to destruction' in which animals are deliberately poisoned to death by raising drug dosage until at least half of a group are killed.

It is impossible to be certain that no unexpected effects will be seen when a new drug is administered to a whole animal even if a battery of non-animal tests has shown that it is probably safe. Thus the first trial of a drug on either a human patient or on an animal is necessarily an experiment, but there is clearly a very great moral difference beween giving moderate doses which all available evidence suggests are likely to be harmless and deliberately raising doses to discover the fatal one. If such a reduction in risk were to be achieved, it is perhaps arguable that human volunteers should be prepared to replace animal 'conscripts' in the final whole-animal tests. However, the question of use of animal volunteers is essentially secondary to the main project of reducing net suffering and death by ensuring that the final whole-animal test is a safety check instead of a poisoning test.

I think it is plausible that many of our relations with domestic animals do involve the violation of tacit contracts of friendship (and that this is why some actions which purely consequentialist considerations would not condemn are in fact wrong). Cows and hens become financially less profitable well before they are physically aged. The egg yield of modern strains of chickens starts to fall after two years, although the hen herself can be fit and active for as much as ten years longer. Cattle are 'culled' from the herd when their udders become pendulous, or damaged by mastitis, or when they become lame because of unsuitable flooring and inadequate care of their feet. It does not seem unreasonable to say that animals who have provided a human being with income all their productive life have some claim upon him when their financial value begins to decline. Battery chickens probably have little relationship with their owners, who may see them for as little as nine minutes per day, but cows definitely do have some kind of relationship of trust with their handlers. If they chose to 'violate' this tacit agreement, then the human partner would be severely injured, but in fact cows sometimes appear to hold back quite 'purposefully' from hurting people. Most dairy cattle lose their calves for sale as veal, only a few being reared as herd replacements, and the cows are killed before physical old age, often in frightening and perhaps painful conditions.[46] It seems that these farming practices must involve something like treachery on the part of humans. Similar considerations must apply to other examples of situations where animals who are in a relationship of trust are exploited, for example, the killing of horses who are no longer capable of heavy work; the dumping of brood bitches who are past the age when they can produce large litters; and so on. It is sometimes claimed (for example, by Paton)[47] that domestic animals benefit compared to wild ones in terms of average life span. In some instances this may be false. Wild turkeys become sexually mature at about 2 years old; domestic hen turkeys mature at about 7.5 months and are then kept for only another six months more (for one egg-laying season). Non-breeding birds are killed at maturity—resulting in 100 per cent premature mortality, something which would be impossible in a stable wild population.[48] Similarly, broiler chickens are killed at about 1 month old, and broiler breeders at 16 months, compared with the wild form's minimum reproductive age of 12 months.[49] However, even if it were true that domestication prolongs the life span of the individual, I believe Paton's argument cannot stand up when it is analysed in terms of personal relationships. The animal does not calculate an actuarial cost/benefit scheme on the basis of which he finds whether or not he is getting a fair deal. He simply trusts the people he has grown up to know. At

---

[46] Farm Animal Welfare Council, *Report on the Welfare of Livestock (Red Meat Animals) at the Time of Slaughter*, Ref. Bk. 248 (London: HMSO, 1984).
[47] W. Paton, *Man and Mouse* (Oxford: Oxford University Press, 1984) 94.
[48] R. D. Crawford, 'Turkey', in Mason (ed.), *Evolution of Domesticated Animals*, p. 326.
[49] Crawford, 'Domestic Fowl', in ibid., p. 298.

every stage he has an expectation that those people will be friendly and will not attack. Unilateral termination of the relationship by the human participants on the grounds that the animal has received all the benefit he is able to 'pay' for is essentially a betrayal. In a pact which was essentially one of mutual trust, non-aggression, and helpfulness, the animal partner has not reneged, it is simply that his means are not what they were. But what the animal 'agreed' to was the formation of a relationship, not a financial bargain. He has fulfilled the 'contract' and will continue to do so. If the relationship were between unequal humans I do not think we should hesitate to say that the cleverer partner has cheated and violated the less intelligent partner's right to have the terms of his contract honoured.

Angela Patmore quotes a captain in the Royal Army Veterinary Corps in her history of the domestic dog: 'In the early fifties a veterinary surgeon came here and removed the scenting membranes of certain dogs [trained to sniff out explosives], yet they could still work. Of course, that doesn't prove anything. Taste and scent are inextricably linked; perhaps they were using taste.'[50] This kind of unprovoked assault by humans seems a clear example of violation of an implicit contract.

Contract-breaking seems, at times, to be 'ritualized', perhaps as a protection against guilt feelings. Successful stockmen tend to be those who develop caring relationships with their animals.[51] In spite of this, it will be these same stockmen who, in the end, betray the animals they care for by herding them to slaughter. Such people very probably *do* feel regret and sorrow when their animals are killed, but I suspect from my own observation of stockmen that this sorrow is prevented from being translated into action by ritualizing emotion in terms of phrases such as 'That's the way life has to be', 'They're not pets', 'People have to eat', and so on. This in spite of the fact that these may be people who, in circumstances not governed

[50] *Your Obedient Servant* (London: Hutchinson, 1984), 95.
[51] M. F. Seabrook, 'The Psychological Interaction between the Stockman and His Animals and Its Influence on Performance of Pigs and Dairy Cows', *Veterinary Record*, 115 (1984), 84–7; Universities Federation for Animal Welfare, *Stockmanship on the Farm*.

by ritual, may be prepared to go to infinite pains, and sometimes to risk even their own lives to safeguard the lives and comfort of their stock. Similarly, individuals who care deeply about animals continue to eat them, not because of greed or self-indulgence, but because they do not dare to risk change. This kind of self-pressure to accept things as they are may be portrayed as a sign of growing up—a surprising number of children's books on growing up involve the 'sacrifice' of a loved animal as a symbol of 'facing reality'.[52] And, similarly, at an institution for disturbed children in America:

Because these urban children are not as accustomed to animals being slaughtered for food as farm children might be, this issue must be addressed caringly. . . . We have been honest and realistic about the production animals' role on the farm, and we have educated the children about where their food and clothing come from. They know not to become attached to a litter of piglets because they will all be sold.[53]

A newspaper report on a British school's agriculture classes quotes the headmistress:

We discourage them [the children] from giving them [the animals] names.[54] . . . This is a very serious attempt to introduce questions of agricultural ethics and to help them come to terms with where their food comes from. It would be wrong to encourage them to treat farm animals as pets. . . . The controversy [over killing animals at the school] arose from a vegan who questioned the whole idea.[55]

The same article reports the reaction of one of the children:

as 14-year-old Danielle Clark, a volunteer responsible for the rabbits, said, nobody likes to think of the animals having to be killed. 'I put up with it,' she added. 'They all have to die sometime.'[56]

In his book *In the Company of Animals*, James Serpell makes the same suggestion that humans have a general tendency to

[52] e.g. Marjorie Kinnon Rawlings, *The Yearling* (London: Pan, 1976); R. N. Peck, *A Day no Pigs would Die* (London: Arrow Books, 1980).

[53] Mary Link, 'Helping Emotionally Disturbed Children Cope with Loss of a Pet', in *Pet Loss and Human Bereavement* (Ames, Ia.: Iowa State University Press, 1984), 82–8.

[54] Cf. the anthropological evidence cited earlier that animals who have received individual personal names are less likely to be seen as food.

[55] The *Guardian*, 23 Mar. 1988, p. 38.

[56] Ibid.

relieve their guilt about harming animals by ritualization.[57] He makes the further very interesting suggestion that one reason why it is difficult to persuade exploiters of animals to adopt moderate reforms is because this would require them to think of animals as conscious fellow creatures (Serpell says, 'as persons rather than things'), and that this would be psychologically painful because it would cause them to feel guilt about harming animals at all (p. 155). He suggests that it is easier for someone to refuse to give *any* consideration to the welfare of animals who are to be killed and eaten, than to be concerned for the animals' happiness and comfort while he is rearing them, but then turn off that concern when the time comes for them to be slaughtered.

Agriculturalists do cheat the system to a certain extent, and a few favoured animals are sometimes reprieved and kept as semi-pets. (For example, I know of one farmer whose herd of beef cattle contains a small nucleus of very tame individuals who are always kept back from the market.) Similarly, James Serpell notes that he met a Welsh sheep farmer who had a couple of elderly ewes who were never sold, and also several researchers who would reserve one or two animals as pets.[58]

Serpell's investigation of anthropological literature revealed a significant pattern. The keeping of cherished pets is very common among tribal hunter-gatherer peoples, tends to be rare among settled agriculturalists, and undergoes a resurgence among urban dwellers. He suggests that the underlying factor in this pattern is the need to avoid guilt about killing familiar animals. Hunter-gatherers kill and eat only wild animals to whom they have no attachments. Similarly, urban dwellers seldom come into contact with the animals they eat, and can avoid making the connection between the meat on their plate and a living, conscious creature. Their friendship with their companion animals does not compromise their ability to consume the anonymous flesh they buy. Agriculturalists, in contrast, live in intimate contact with the animals they are ultimately responsible for killing and they simply cannot afford to see those animals as persons. Serpell suggests that it was for precisely this reason that Western theological and

[57] *In the Company of Animals*, pp. 136–70.
[58] Ibid., p. 157.

philosophical systems tended to insist that animals were soul-less, worthless, or without consciousness until our civilization had begun to develop substitutes for the exploitation of animals and an urban population had begun to develop.[59]

Accepting that animals have moral status and/or rights cannot be expected to eliminate moral dilemmas about the way we should act towards them. We shall still often not know which course is the right one or be tempted to rationalize our own self-interest.

In practice human beings have *de facto* control of the lives of both wild and tame animals to an extent which cannot readily be relinquished, except perhaps in the comparatively untouched oceanic wilderness. Allowing wild deer to multiply until they starve in countries where humans have eliminated predators is not respectful of their rights. The very completeness of human control of domestic animals, which so offends many who genuinely wish for the welfare of animals, means that reforms can be more easily implemented. If there were the will to do it we could easily ensure that the population of cats, dogs, and horses was controlled by manipulation of numbers born instead of by slaughter; similar control of deer or elephants is a more remote (thought not impossible) prospect. Dogs can live healthy lives on a vegetarian diet and while cats require some animal fat and protein I see no reason why a reformed society should not organize matters so that they performed a natural scavenging role, consuming large animals who had lived out their natural spans. In fact in Britain pets are already essentially scavengers on the leavings of human meat-consumers, since pet food is made from meat and fish waste.[60] We need to stop seeing ourselves as completely alien to the rest of nature. Living as a domestic animal is not really any more 'unnatural' than commensalism with humans (for example, rats, mice, and sparrows); scavenging after larger animal predators (vultures); or enjoying a relationship of mutualism, like that existing between large tropical-reef fish and cleaners, smaller fish who remove and eat harmful parasites.[61] The only important

[59] *In the Company of Ánimals*, pp. 48–58, 136–8.

[60] Tim Potter, *British Animal and Pet Food Companies* (London: Jordan, 1984), 11.

[61] For a fuller discussion of mutualism between different species of

difference is that humans have a responsibility to act justly towards moral patients within their community. They can of course refuse to accept this responsibility, with morally bad consequences, but they can also choose to accept it, together with the promise of something better then the amoral cruelty of nature.

Opponents of the concept of animal rights sometimes attempt to base their arguments upon the achievements of humans. However, leaving aside the fact that a considerable proportion of these achievements is of comparatively recent origin (there is a good deal less of a gap between the abilities of a band of chimpanzees and those of palaeolithic man than between the same chimpanzees and space-age man), we must recognize that the staggering achievements of modern humans are social phenomena. Since domestic animals form a part of the social structure of human society they necessarily share in the achievements of that society. Indeed, it is a basic claim of the defenders of unrestricted medical research that modern medical achievements depend almost entirely upon the use of animals. Such animals may perhaps be largely regarded as depersonalized 'tools', but at least some animal research does depend upon a degree of co-operation from tame animal subjects. It might be objected that no non-human animal is able to understand or directly use human discoveries, but it is equally true that no human is capable of understanding or controlling the whole of modern science, technology, or art. Humans have the power to destroy the Earth, control the genetic basis of life, and investigate the very earliest moments of the universe, but most of us are not *personally* capable of any of these things. There is a sense, therefore, in which socially co-operating domestic animals share these powers in the same way as human members of society. It is surely no accident that these are also the animals who are most likely to be seen as 'quasi-persons' in Ingold's sense.

The anthropologist Edmund Leach points out that the concept of *people* is essentially a social one.[62] 'Primitive' groups of humans have a tendency to group 'friends' and

sea animals see L. P. Zann, *Living Together in the Sea* (Hong Kong: T.F.H. Publications, 1980).

[62] *Social Anthropology* (Glasgow: Fontana, 1982), 55–85.

'enemies' together as 'people', in opposition to 'strangers', who are not people and may be exterminated virtually at will. Only 'people' can be enemies or potential partners, and it is only in wars against people that one feels any need for limits on the way in which opponents may legitimately be treated. I suspect that this idea of the social determination of who counts as people might be more helpful in considerations of the status of non-humans than attempts to draw up a check-list of qualities or capacities which are necessary conditions of the status of a *person*.[63] For example, it avoids the counter-intuitive tendency of such lists to exclude severely defective humans, since these are necessarily classified as people in virtue of their membership of human families and their possession of human parents. Our ability to assess who can be classified as 'people' perhaps involves something more like a Gestalt perception than the performance of a scientific exercise in taxonomy, with a variety of different factors being simultaneously weighed. It is significant that 'animal lovers' do have a tendency to use the words 'person' and 'people' as terms of endearment when addressing or discussing their companion animals, and I suspect that a straw poll would indicate that they see animals as people (but *not* as men).

The question of whether animals should be classified as a type of person seems to be less important to a consideration of their moral status than that of their ability to feel and to foresee pain and other harms. I suspect that the significance which some philosophers have attached to the status of persons[64] is partly due to the importance of legal personhood to the question of whether animals can be said to have rights in law. As Paul Matthews notes in his two articles on property left for the support of individual animals,[65] there are clear disadvantages to animals in their present legal status as mere objects. People who have companion animals often wish to

[63] For a discussion of the concept of a person see J. Teichman, 'The Definition of *Person*', *Philosophy*, 60 (1985), 175–85.

[64] e.g. P. Singer, *Practical Ethics* (Cambridge: Cambridge University Press, 1979), 78–90, 93–105.

[65] Paul Matthews, 'Trusts to Maintain Animals', *The Law Society's Gazette*, 80 (Oct. 1983), 2451–2; and 'Maintaining Pet Animals after Your Death', *The Cat* (Journal of the Cats' Protection League), 57 (Mar./Apr. 1984), 12.

make wills in which money is left in trust for the animals' support. However, it is not legally possible in English law to leave money to an animal (who is himself an owned chattel). Strictly speaking, it is not legally valid *either* to leave money to the animal direct *or* to attempt to set up a trust for his maintenance (since trusts are supposed to have a person as their beneficiary). Trusts to maintain animals (and certain inanimate objects, such as graves) have been accepted by the courts in the past on certain occasions, but Matthews considers that it would be possible for relatives of the deceased person successfully to challenge their validity on the grounds that the beneficiary of a trust must be a person. It is possible to get round this to some extent by making a will in which money is left to a human on condition that the animal is cared for, but the animal still lacks the kind of protection given (for example) to a child who is left money in trust since the child is a legal person and a lawsuit can be brought on his behalf if he is defrauded. Further legal complications arise because trusts cannot be of indefinite or perpetual duration (for example, the perpetual maintenance of a particular grave), but must *either* be for a fixed time span (English common law probably limits this to a maximum of twenty-one years, although there is some uncertainty on this point), *or* for the duration of the life or lives of specified humans who are alive at the time of the making of the will. It is not possible to 'tie' the duration of the trust to the life of the animal it is designed to benefit. This might not be very important for short-lived species like dogs and cats who are unlikely to survive the testator by more than twenty-one years, but it has obvious significance in certain cases (Matthews mentions one trust which was set up to benefit a parrot). Furthermore, the animal in question is himself part of the testator's estate so that unless the will specifies who will be his new owner, a situation could occur in which the testator's next of kin would inherit him, and could then claim the funds of the trust intended for his benefit. This possibility arises because any fund of money set up to benefit the property of another must itself be the property of the latter, without any necessity to apply it in the way specified in the will.[66]

[66] In a recent case in the USA, the trustees of an estate left for the benefit of a collection of monkeys and apes used the trust funds to set up a research

Similarly, legal personhood would mean that anti-cruelty laws would cease to appear to be more like laws of manners than the protection of individuals and it would no longer be possible to equate cruelty with 'victimless crimes', as Lord Patrick Devlin does in his book on the enforcement of morality.[67] Since the justifiability of legislating against behaviour purely because we happen to disapprove of it, rather than in order to protect its victims, is highly controversial, it appears that the legal recognition of animals as persons, i.e. as individuals who deserve protection for their own sakes, could be an important safeguard for their continued legal protection. It would also pave the way for further improvements. Proposals for the ending of ritual slaughter in Britain, for example, have been strongly criticized as mere prejudice against the customs of minorities:

If ritual slaughter is cruel . . . it is for the minority communities themselves to sort the matter out according to their own customs and priorities. Members of the majority community have no right to lecture them.[68]

In ritual slaughter according to Jewish and Islamic custom pre-stunning is not permitted and the animal's throat is cut while he is still conscious. There is evidence that the cut does not produce immediate loss of consciousness, particularly in young animals, and that there may be a period of severe pain.[69] If animal protection were not seen primarily as a question of cultural norms, but as an attempt to protect a particular group of society members, reformed legislation need not be seen simply as prejudiced interference with minority culture.

The 'personhood' of some animal species need not mean that *all* animals would have to be granted equal status. As I have noted previously, some species, such as sponges, are unlikely to have subjective experience. Nor need it mean that animal persons must necessarily have exactly the same legal rights as

laboratory conducting experiments on primates (*Newsletter of the International Primate Protection League*, 11(2) (1984), 11–12).

[67] P. Devlin, *The Enforcement of Morals* (London: Oxford University Press, 1965), 17.

[68] N. Redfern, letter published in the *Guardian*, 2 Sept. 1985.

[69] Farm Animal Welfare Council, *Report on the Welfare of Livestock when Slaughtered by Religious Methods* (London: HMSO, 1985).

normal adult humans. In some instances this might not be in the animals' own best interests, as in the case of euthanasia, where it is arguable that the animal's inability to comprehend his suffering is a crucial reason why it should be permissible to end his life painlessly if there is no reasonable hope of his recovery. However, it does not seem to be the case that present laws necessarily compel us to treat different types of legal persons as if they were exactly the same. No one would treat the dissolution of a limited company as murder, even though companies are treated as legal persons in certain circumstances. Thus, legal classification of some species of animal as persons need not rule that their interests might *never* be sacrificed to the welfare of the greater community (as indeed the interests of human persons are sometimes sacrificed in times of great necessity). However, we might no longer feel so easily justified in killing them, as the Nuer say, 'just for nothing' (i.e. merely to eat, not out of necessity).[70]

[70] Evans-Pritchard, *Nuer Religion*, p. 265.

GLOSSARY

Alga Primitive, photosynthetic eukaryote (q.v.).

Altruistic A technical, biological term for any action which has the effect of benefiting other entities at the expense of the acting entity.

Animalia The Animal Kingdom. Multicellular, eukaryotic (q.v.) organisms. Never photosynthetic. Generally motile at some stage of life history.

Ape Group of large, tailless primates, closely related to humans. Sometimes also used for certain species of monkey.

Australopithecus Lit. 'southern ape'. Extinct group with affinities to both the living great apes and humans.

Bacteria Single-celled prokaryotes (q.v.).

Cephalopoda Group of marine molluscs notable for their generally large size and highly developed nervous and sensory systems.

Commensalism An ecological relationship in which one species lives in close association with another, often making use of discarded food, etc., but causing no significant harm or benefit.

Elephas Generic name of the Indian elephant.

Euglenids Group of photosynthetic protozoa (q.v.).

Eukaryote Single or multicellular organism possessing a membrane-bound nucleus.

Fungi Non-photosynthetic eukaryotes (q.v.), most of which are non-motile and plant-like, although some members (the slime moulds, q.v.) form motile, amoeboid cell masses, which feed by ingesting organic particles.

Gorilla	Generic and species name of the gorilla species.
Great apes	Chimpanzee, gorilla, and orang-utan.
Homo	Generic name of the human species.
Loxodonta	Generic name of the African elephant.
Mesozoa	Group of small, multicellular parasites lacking specialized sense organs or nerves.
Metazoa	Multicellular, generally motile, non-photosynthetic organisms other than fungi.
Monkey	Group of medium-sized primates possessing tails. Closely related to humans, but less so than the apes (q.v.).
Mutualism	A close relationship between two species in which each benefits the other.
Pan	Generic name of chimpanzee species.
Plantae	The Kingdom of green plants. Multicellular eukaryotes (q.v.) which are typically photosynthetic, but may secondarily lack chlorophyll.
Pongo	Generic name of the orang-utan.
Prokaryote	Simple, single-celled or loosely colonial organism lacking a membrane-bound nucleus.
Protozoa	Single-celled, non-photosynthetic, motile organisms.
Selfish	A technical, biological term for any action which has the effect of benefiting the acting entity at the expense of other entities.
Sepia	Generic name of cuttlefish species. Member of the class Cephalopoda.
Slime mould	Motile eukaryotes (q.v.), normally classified with the fungi. Non-photosynthetic, amoeboid, possess single-celled and aggregated phases.
Virus	Non-cellular, obligate parasite of cellular organisms, little more than protein-coated nucleic acid.
Volvox	A protozoan which characteristically lives in clusters or colonies.

INDEX

0486